4.04

Essay Index

D1505534

UNLESS
PEACE
COMES

UNLESS
PEACE
COMES

A SCIENTIFIC FORECAST
OF NEW WEAPONS

EDITED BY

Nigel Calder

Essay Index

THE VIKING PRESS · NEW YORK

CONTENTS

INTRODUCTION

ABDUS SALAM
Pakistan

*Professor Salam is director of the International Centre for Theo-
retical Physics, Trieste, and professor at Imperial College, London. He was elected a Fellow of the Royal Society in 1959. He
is Scientific Adviser to the President of Pakistan.*

ON A COLD February morning in A.D. 1258 the greatest living
astronomer of the age, Nasir al-Din Toosi, was roused on the
orders of his Mongol master, Hulagu Khan, the grandson of
Genghis. The Mongols' siege guns were positioned outside
the city walls of Baghdad, and Hulagu had planned to give
the signal of attack. But that night he had received from the
reigning Caliph of Baghdad, al-Mustasim, a dire warning:
that if Baghdad were attacked, the Caliph—as the last suc-
cessor of the Prophet of Islam—had been vouchsafed succor
with such weapons of mass destruction as the world had
never seen before. Sheets of fire would rain from the heavens,
black death would smite the Mongol armies, and the sun

would darken, never to shine again. The Mongol's pagan fears were roused; he was in a state of terror.

Toosi the astronomer was no mean scientist. His are the first recorded doubts about Euclid's postulate of parallels; in his work occurs the first foreshadowing of non-Euclidean geometry. He was a devout Moslem, but he considered the reigning Caliph a usurper. Toosi could not offer protection from the fire from heaven, should any come, nor from the incidence of black death, but he did assure his Mongol master that no one could interfere with the course of the sun. He said he would willingly forfeit his life for all it would then fetch, if the Caliph proved right.

Hulagu gave the word for attack, and Baghdad was stormed. Over a million men, women, and children were massacred that day by blood- and fear-crazed Mongols; among them was the Caliph al-Mustasim. With the Caliph died the 600-year-old institution of the Caliphate in Islam, as also died the even older tradition of Arabic learning and scholarship, never to recover again.

The pages that follow in this volume speak of weapons of mass destruction, either already in existence or capable of development. They read like the letter that the ill-fated Caliph sent to the Mongol. Although Toosi's name has ever since been cursed and reviled in all the Arab lands, for once I wish one could offer, like him, one's life in forfeit on the gamble that such weapons would never be deployed. Unfortunately, if the past is any guide, the contrary is more likely to happen; barring any moral miracles, one can only expect that, however fanciful some of the horrorsome anticipations of the succeeding pages appear today, these and worse may indeed be realized.

Much of the book is concerned with weapons of mass destruction and their delivery systems. These include atomic, biological, and chemical weapons, with the ultimate sophistication of geophysical weapons yet to come. Of these, the destructive power of nuclear weapons is best publicized. Paradoxically, this very publicity has bred false security: "the world and its statesmen have become inured to 'living with the bomb,' " as Dr. Inglis remarks. Few realize how slender is the basis of the present nuclear stalemate; how the seemingly inevitable nuclear proliferation of years ahead and, worse still, the partial defence offered by antiballistic missiles are likely to erode even this tenuous basis. Less well known than the nuclear is the awesome destructive potency of the chemical weapons already in existence—nerve gases, incapacitants, psychic poisons. Still less appreciated is the progress made in recent years with aerosol spraying of deadly biological agents, and very few of us indeed have suspected the possibilities of tampering with the geophysical environment —for example, the burning of a selected territory on earth with intense ultraviolet rays from the sun, by puncturing holes in the naturally protective ozone layer above.

Add to this the ever-increasing deadliness of conventional fire power through the magic use of electronics; the ever-increasing sophistication of interception and counterinterception systems. Add, too, the possible perfecting of self-controlling, self-propelling robot bombs. It is clear that this fast-shrinking earth will soon be very, very narrow indeed for the hunted sons of Adam.

We are more than indebted to Nigel Calder for persuading his distinguished contributors to write. But what can a book like this hope to achieve? A revulsion against weapons of mass destruction? A turning away from all warfare toward a juridical settlement of disputes? An urge toward complete disarmament? Perhaps hasten the withering of the concept of the nation state—the concept that unfortunately underlay the United Nations and has all but wrecked it, the concept that stills for most of us the protests of our humane consciences when we create, design, and fabricate those weapons of mass murder?

Perhaps we shall help to restore the idealism, the vision of a mere six years ago, when the United Nations Disarmament Committees could solemnly assume that, with coming disarmament, an annual $120 billion (and, even more valuable, the scientific resources of the world) might be released to eliminate poverty, hunger, and disease for all humanity? The fear of nuclear annihilation has so far proved too feeble to achieve this. The fear of biological weapons may be more potent. For one thing most of us have experienced in our own lives, some time or other, the despair of lingering disease, which somehow is even more a part of the human condition than death. For another—in Professor Hedén's words— "Man's success with biological control in insects and other pests indicates a road which is certainly appalling. . . . The war gamesmen who balance the cost of anti-ballistic missiles against 'acceptable' megadeaths seem to walk on thin ice, in presuming fair play or Expected Gentlemanly Behaviour (EGB) by opponents as far as biological weapons are concerned."

The complaint is often heard that scientists fail to alert their fellow citizens to possible consequences of current developments. In this book scientists of six countries and several fields of research have joined together to give clear warnings. I hope their words do not go unheeded, that no fire rains on earth and no black death smites the human race.

UNLESS
PEACE
COMES

BATTLEFIELDS OF THE 1980s

BY

ANDRÉ BEAUFRE
France

Général d'Armée Beaufre is director of the French Institute of Strategic Studies, Paris. At the culmination of a distinguished military career, he served as Joint Chief of Staff at the Supreme Headquarters Allied Powers (Europe).

THE NOTION of the *guerre classique*, or "conventional war," corresponds to the idea of a military conflict wherein the combatants refrain from using the new "scientific" weapons of mass destruction, whether atomic, biological, or chemical. So defined, conventional warfare complies with a wish to use less destructive weapons; it would therefore amount to a voluntary humanization of war. This self-imposed limitation can also result from the wish not to bring into play new means of combat whose military and political consequences are poorly understood. As we shall see later,

1

this limitation will often result from inhibitions produced by nuclear deterrence and the risks of escalation, as well as by the pressure of world opinion. Nevertheless, to consider conventional warfare as being humane could in many circumstances prove dangerously wrong as the risk of catastrophic destruction becomes even greater. I shall return to this crucial question of the humanization of warfare in my conclusion.

To review the possibilities in conventional warfare in the 1980s, I shall examine first the prerequisites for conventional war to recur within the strategic framework determined by the existence of nuclear weapons, then the technical characteristics that one can envisage for future conventional warfare, and finally the operational features of these wars as we can imagine them today.

The Scope for Conventional Warfare

The existence of nuclear weapons has created altogether novel dangers. On the one hand the risk of destruction entailed by war is now of an order of magnitude quite disproportionate to our experience. A medium-sized thermonuclear bomb of one megaton produces an explosive force equal to a salvo of 200 million "75s"—the gun that was "modern" in 1914! Not only are the destructive effects of a nuclear war difficult to compute rationally, but also there is, so far, no reliable means of self-defence. In these conditions, nuclear war involves the combatants in the certainty of mutual destruction, which completely dispels any adventurous illusions about the outcome of the conflict. On this certainty rests the effect that the nuclear deterrent imposes on the use even of conventional weapons.

But this nuclear deterrence is neither absolute nor spread evenly over the whole globe. The enormous risks of nuclear warfare make it an implausible threat in minor conflicts. Moreover, in areas of the world where the interests of the nuclear powers are not vital, nuclear deterrence does not apply—for instance, in Vietnam at the time of writing. Even in the regions where these interests are vital, it is difficult to imagine that nuclear war could be unleashed in response to local incidents of secondary importance. In the present strategic nuclear situation, therefore, the normal field of action of conventional war is in those parts of the world not covered by nuclear deterrence—virtually the whole world outside the United States, the Soviet Union, and Europe. Furthermore, the possibility of very limited and doubtless very brief conventional military actions cannot be completely excluded in the zones covered by nuclear deterrence, for example in Berlin or in Eastern Germany. But such confrontations should not normally become more than local incidents if the risk of escalation to the nuclear level is to be avoided.

Nevertheless, the present strategic nuclear situation can be modified, either by the effect of innovations such as antiballistic missiles, or by disarmament agreements that neutralize the nuclear weapons. In this context, the possibility of minor or major conventional wars could reappear in Europe as the result of political events.

These different possibilities involve widely different kinds of conventional conflicts. If it is a matter of taking advantage of a favorable situation to achieve an object in spite of the existence of the nuclear deterrent, the war will take the form of a swift military action designed to achieve some kind of *fait accompli*, as did the 1956 Israeli campaign in the Sinai Desert. If, on the other hand, the conflict takes place outside the comparatively small zone affected by nuclear deterrence, conventional warfare can manifest itself in more or less important military interventions, of which recent examples are not lacking—whether to occupy or defend a country against invasion (Korea), to uphold a government against subversion (Vietnam, Yemen, the Dominican Republic), or to defend national interests (Suez).

Finally, suppose that in Europe or on the Chinese frontier, the nuclear deterrent came to lose its power. Then we might witness the development of classical military conflicts gradually increasing toward major importance until they regain the intensity of the World Wars of this century, wherein the nations exhausted their strength in attrition and threw everything they had into the fight. This is fortunately not the most likely hypothesis, since the nuclear deterrent will retain enough of its menace to limit the scope of warfare for a long time to come. But it is a very dangerous possibility. World Wars I and II, waged only with conventional weapons, caused great enough turmoil in Europe and the world; another great war involving the use of conventional weapons of the future could be immeasurably more devastating. Certainly, humanity is still sensitized by its recent terrible experiences, but memories fade fast from one generation to the next. In the 1980s, the coincidence of a reckless head of state and a people inflamed by a great passion—racial, nationalist, or ideological—could reproduce the phenomena that twice in thirty years devastated the world and led Europe to her eclipse.

General Trends in Conventional Weapons

These anticipations lead one to consider possible progress in the realm of conventional weapons. They have developed spectacularly since the beginning of the century, but since World War II the process has been accelerated by the scientific discoveries and technical innovations of all kinds that characterize our times. Speeds and ranges, for example, have increased in an extraordinary way since 1945, as has the effectiveness of military telecommunication. Nevertheless, one must acknowledge that, in general, conventional weapons had already evolved remarkably in the course of the two World Wars, and although one can foresee important new technical improvements, there may be no revolutionary change in the types of weapons between now and the 1980s. But, as we shall see, this limitation does not preclude far-reaching changes in the operational conduct of conventional war, brought about by evolutionary improvements in existing weapons.

From the technical point of view, one can generalize to say that existing weaponry will be considerably improved by the use of new materials and new methods of manufacture and that their effectiveness will significantly increase with a better use of electronics.

Present-day materials will be partly replaced by composite materials. Plastics, in particular, will no doubt acquire mechanical properties, coefficients of expansion, resistance to high temperatures, heat conductivity, and so on, which will allow their use jointly with reinforcing metals to form the structural materials for vehicles, for armor-plating, and even for guns. Such materials will give great savings in weight; they may also achieve, for armor-plating or structures, a resilience which will increase their resistance to damage. The materials may also be nonreflective to radar.

New techniques of production, using computer-controlled machine tools, or explosive, electrolytic, and magnetic forming of the workpieces, can greatly reduce manufacturing costs of weapons and also allow wider use of "difficult" materials.

But it is above all by the military application of advances in electronics that we shall be able to enhance the efficiency of weapons. The pinpointing of targets by land-

based, airborne, or satellite radar, the use of infrared to reduce the concealment obtained from darkness and overcast
weather, and miniaturized battlefield computers will together allow for a centralized control of conventional firepower; its efficiency will be further increased by the use of
advanced proximity fuses that detect their targets. Such
technical progress will put air defence on a different footing
and reinforce its value. Again, the military use of satellites
cannot help extending the province of air defence to
space itself. Most of these technical trends lead one to envisage more and more complex, and therefore expensive,
weapons, and hence a gradual reduction in the size of military forces, though with a corresponding increase in firepower and mobility.

Particular Weapons of the Future

If we now pass on to examine each category of weapons, we shall meet these general considerations again, but
with significant differences between the categories.

The means of conventional antipersonnel fire—the rifle,
machine gun, mortar, and artillery—seem to have attained
their definitive forms. Certainly there will be an improvement in weight, range, precision, and rates of firing, but the
limitation of these killing machines will always be the supply of ammunition. The greater their consumption of shot,
the more necessary it becomes to avoid the blind shooting
at zones that was in vogue during World War I. Interest
now concentrates, as I have indicated, on the means of pinpointing targets and on the centralized control of long-
range shots. If, as one may well think, there is substantial
progress in this respect, it is possible that all movement of
troops in the open could be countered by fire over a great
depth of the battlefield. With high-fragmentation shells,

detonated at the proper height by proximity fuses, infantry movements could become very hazardous or even impossible. Two solutions could be attempted. One would be the organization of a great many simultaneous movements, since it would be impossible to direct fire everywhere at once; even so, losses would be terrible. Alternatively, one can look to the general use of lightly armored vehicles giving protection against antipersonnel weapons. Movements would be speeded up and would involve mechanized infantry similar to that considered tactically necessary in nuclear warfare.

This last conclusion focuses attention on weapons systems conceived as a function of tank warfare. Here, one can foresee some important changes, for the evolution of the battle tank is far from complete. The spread of teleguided antitank weapons, with shaped charges giving high penetration, will render heavy armor-plating and turrets more and more useless. The tank-mounted antitank gun, with its high muzzle velocity and large caliber, is itself being outclassed by teleguided missiles. These two considerations, combined with the anticipated progress in the quality and elasticity of armor-plating, force us to think of much lighter designs, probably amphibious, which will lead to great changes in the defensive value of the terrain. Nevertheless, the tank will remain helpless in the face of thick forests and steep escarpments; these two will continue to be the chosen terrain of infantry on foot.

The lightening of the tank will have another important consequence: that of making it readily transportable by air. Airlifts of powerful weapons will lend greatly increased speed and range, and hence effectiveness, to military operations. But here we encounter difficult problems whose solution can be very different according to the weapons systems selected for development. As I have already noted, there will probably be a great improvement in aerial interception and missiles, especially because of their relevance to nuclear warfare, so that the use of aircraft, and of transport aircraft in particular, may become most problematical. It would therefore be only after the enemy's air defences had been disrupted that airborne operations could be envisaged. The preliminary air battle aiming at the "mastery of the air," the great lesson of 1945, would thus be replaced by a preliminary battle for the "freedom of the air," presumably based on missiles armed with conventional explosives. The manned aircraft would then be no more than a means of exploitation once "freedom" was achieved.

On the other hand, one can also imagine that countermeasures against interception based on a combination of electronic means of jamming enemy radar, missile attack, and the use of hedge-hopping aircraft could permit a reasonable survival of aircraft in flight at least in one sector and for a given time. In such circumstances not only would the combat aircraft retain its present role, but air transport of ground forces would be possible from the start of operations. Naturally, it is impossible to tell in advance if this hypothesis would be born out or, above all, if the presumed efficiency of the means of interception and counter-interception will be satisfactory. The truly enormous experiment that will take place at the beginning of a conflict will be decisive in this field and could have very surprising results. It will therefore be necessary to know how to adapt quickly to the proven realities of the air battle.

It is within this new framework, the uncertainties and constraints of which cannot be exaggerated, that flying machines will have to develop: missiles, aircraft, helicopters, even earthborne vehicles capable of short jumps into the air. Appropriate technical progress is already under way, with short- or vertical-takeoff aircraft, variable-geometry wings permitting a wide range of speeds (maximum speeds of several times that of sound), swift transport planes of large capacity, a spectrum of helicopters from the very light to the crane helicopters, and so on. . . . All these machines will only find their proper roles once the problems of interception and counterinterception have been resolved; that is the pivot on which everything else turns.

From this point of view the satellite can play a decisive part in surveillance and communications. According to the technical systems achieved, it can give the advantage either to interception or to counterinterception. In either case it necessarily provokes the opening of a theater of operations in space, where each adversary will seek to eliminate enemy satellites, with missiles or with manned satellites. Possibly this stage may not be reached by the 1980s, but come it must, for it corresponds to a logical imperative: mastery of space will determine mastery of the air, just as mastery of the air will largely determine the options on land and sea.

In the naval sphere much the same considerations apply as on land. Doubtless improved location and fire-control techniques will make surface navigation very risky until interception has reached a reasonable level of efficiency. Just as the fleets of the 1940s had to adapt quickly to anti-aircraft defence, so the fleets of the 1980s will have to be able to protect themselves against missiles. The evolution that began with gun-carrying ships and continued through aircraft carriers now leads to the development of missile-carrying ships. But two novel features of naval warfare must be emphasized: the considerable increase in speed and range achieved with nuclear propulsion, and the capacity for survival of modern submarines.

By the 1980s widespread use of nuclear propulsion in warships is certain, just as fuel oil dethroned coal, which earlier displaced sails. In new forms, submarines and surface ships will have a completely new mobility. With their evasion potential and transport capacity increased, they could compete with air transport in support of land operations.

As for the submarine, its true characteristics have emerged only since the development of nuclear-powered boats for carrying nuclear missiles. The modern submarine is virtually undetectable when submerged, because of the physical properties of sea water which make detection pulses difficult to propagate. Possibly by the 1980s someone may have discovered a practical means of tracking down submarines at depth, but nothing is less certain. And it is also possible that such new techniques as plastic antiradar hulls and silent motors could postpone the moment when submarines would again become vulnerable. In these conditions it is plausible that fleets will tend to become more and more subsurface, with missile-carrying submarines, hunter submarines, and even transport submarines. For this evolution, leading to the concept of a "war of the ocean depths," the interval until 1984 is perhaps somewhat short, but such an outcome is inevitable if the arms race continues at its present rate.

Before leaving these technical trends, I think it is necessary to reiterate the capital role of electronics. In tracking, centralized fire control, interception, counterinterception and one's consequent capacity or incapacity for movement by ground, sea and air—everything will depend on the efficiency of the electronic systems. These are still largely conjectural. By suitable progress, notably with lasers, the problems of jamming can be overcome. On the other hand, techniques might be developed that would permit the blinding of a part or the whole of a system. In short, the age-old duel of guns and armor may give way to a duel of electronics and counterelectronics.

Operational Realities

In the preceding sections I have considered the technical possibilities confronting two comparable adversaries, where the relative value of the new techniques is uncertain. This situation can apply only where two opposing countries are devoting an important part of their national revenue to armaments (France and Germany in 1914 and 1939, the U.S.S.R. and the United States in 1967). Without saying that such a situation is improbable, one should note that it does not correspond to the general rule, because most normal wars involve forces of unequal development. Indeed, the hope of military success by the more advanced adversary is what makes conventional warfare most likely to break out. It is therefore important to examine the different forms that conventional warfare may assume, according to whether the two opponents are technically unequal, very unequal, or fairly well matched. On this basis one can foresee three main types of war: the war to establish a *fait accompli*, the counterguerrilla war, and the conventional "great war." Naturally, intermediate types and combinations are possible, but these three types are sufficient for purposes of analysis.

The war to establish a *fait accompli* corresponds to the case where one of the adversaries has a marked technical and military superiority over the other and takes advantage of this superiority to secure a political objective without too much risk of setting in motion the mechanism of nuclear escalation or even of psychological escalation in world opinion. To do so, the aggressor must attain his chosen objective quickly, if possible in a day or two, or at the very most within a week. Unless the objective is of modest dimensions (Berlin, for example, where the tactics would simply involve armored vehicles), such a result can normally be achieved only by the use of air transport. That presupposes that the "freedom of the air" is the aggressor's, with a definite superiority in the technical field and in counterinterception, as discussed earlier. It also assumes that the aggressor has plenty of air transport capable of ferrying infantry and vehicles (bicycles at least!) as well as good support by air-to-ground fire. Such an operation is easy to stage from adjacent territory, but if it has to be conducted from a distance, warships carrying aircraft and helicopters are generally indispensable. Ships of this type, although likely to be useless against a modern and versatile enemy defence system, will continue to play an important part in expeditions against secondary military powers. This general outline is based on the lessons of recent actions. Those that were successful (the Israeli campaigns in the Sinai, Soviet intervention in Hungary, American action in Lebanon and the Dominican Republic) achieved the required condition of speed. The ones that failed (the North Korean attack, the Franco-British campaign at Suez) did so essentially because of their slow tempo, which permitted a paralyzing reaction to set in—military in the case of Korea and political in the case of Suez. Presumably the future holds yet more examples of the same rule.

Counterguerrilla warfare is unhappily very familiar as the frequent accompaniment of the process of decolonization. Despite its repetition in many forms, it seems that the character of this kind of war is still not properly understood, mainly for want of an over-all view setting the military operations in a political frame suited to the situation. As the Americans have again recently shown in Vietnam, an overwhelming material superiority of modern military forces over the guerrillas tends to foster policy based on purely military decisions, when in such a case, above all others, the psychological character of the struggle makes the political circumstances supremely important.

But what amounts to an important innovation in recent counterguerrilla wars comes from the considerable improvement in guerrilla tactics and strategy, under the Soviet and Chinese influence. In learning systematically to refuse to fight whenever he is not in a markedly superior situation, and in operating a highly coordinated system of dispersal, infiltration, and terrorism, the guerrilla has acquired a miraculous capacity for survival and for maintaining his hold over his people. In this situation, conventional forces are in the position of a lion attacked by mosquitoes; they can go anywhere they like, almost without a fight, and they can defend the points they occupy in force, but because of the extent of the terrain and its natural obstacles they cannot be everywhere at once. Vast areas will thus necessarily escape their permanent control, and there the guerrilla can persist. As a result this kind of conflict can last for years, often finishing with a compromise born of weariness in the more powerful adversary. Such a pattern occurred in the Algerian war, and it may repeat and grow in the future.

To put down the guerrillas, the most powerful conventional weapons have been tried in Vietnam; in the future perhaps even more modern weapons specially designed for the purpose will be brought into use. It is certain, for example, that developments in air-to-ground fire, the great tactical and logistic mobility made possible by the helicopter, day-and-night surveillance of the terrain by radar, the use of electronic or chemical methods to "label" the friendly population and the guerrillas in some discreet way, and the use of chemical incapacitants will give military forces increasingly effective means of action. But the essential difficulty in guerrilla warfare still stems from the physical impossibility of simultaneously controlling the whole terrain, even with greater and greater resources. That is why psychological methods are most often the only decisive ones. By contrast, the use of all these increasingly powerful modern weapons can mean subjecting the general population to almost inhuman stresses. With these massive and often blind techniques, counterguerrilla warfare comes to look more and more like a war against the population, the opposite of its declared objective. Personally, I believe that more subtle methods—and more political ones, as I have mentioned—are the only ones that will achieve results. Failing that, counterguerrilla warfare is simply an extremely costly means of postponing political solutions, which will impose themselves sooner or later.

The idea of a conventional war between adversaries of technically comparable strength is luckily not very realistic. As I have said, this situation could result only from a conjunction of grave political disagreement between two developed nations and the neutralization of nuclear deterrence. If these two conditions coincided, then we should witness the rebirth of the conventional "great war." What form such a conflict would take in the future is extremely conjectural, for it would depend directly on the relative value of the technical systems favored by the two sides. According to circumstance, one might be faced with a *Blitzkrieg* entailing the quick and complete defeat of one side, or with a prolonged war resolved by the exhaustion of the weaker or less determined party, or ending with nuclear escalation or some kind of revolutionary war.

Blitzkrieg (lightning war) in modern guise presupposes that one of the opposing parties has managed to gain complete ascendancy in power and mobility while driving his opponent out of the air. The touchstone here is the right solution of the counterinterception problem. If this result were achieved, the side that benefited would possess extraordinary freedom of maneuver because the volume of modern forces, increasingly limited by their great cost, could not adequately defend an extended theater of operations against an attack in depth. The campaign would then take the shape, already indicated, of a war to establish a *fait accompli*, after a short preliminary phase dedicated to disruption of the enemy's interception systems. Vast airborne penetrations would reproduce, in quicker tempo and greater force, the sequence of events seen in France in 1940 following the invasion by German armor. Naturally, the idea of a "front" would lose all practical value, as would narrow seas like the English Channel.

If, on the other hand, the interception systems on both sides function effectively, airborne movements become impossible. If both sides' fire power and control are also well matched, there might arise a kind of stabilization of fronts formed by dispersed and mobile forces. In that case, well-organized thrusts by armor might produce a decision, but it would be a much slower process. If, meanwhile, antitank defence had made qualitative and quantitative progress, a completely stabilized situation might develop involving the progressive mobilization of armies of foot soldiers dedicated to maintaining the continuity of the defence front. Conventional war would then revert toward the techniques of attrition, where the demographic and industrial power of states would come into play, as in the two World Wars. Destruction of life and property would then be considerable—much greater than in the previous conflicts, because of the improved power of the military machine. Victory would finally belong to the party that managed to re-establish, in its own favor, conditions for great operational mobility. Failing that, there would be a strong temptation to have recourse to a surprise nuclear attack to obtain a decisive success. The conventional war would be over, but with all the grave consequences entailed by nuclear escalation.

More normally, such a prolonged trial of strength could not fail to give rise to revolution. As in Russia in 1917 and Germany in 1918, military defence could be ruined from within and bring with it the downfall of the political and social system.

Conclusion

This brief review of the prospects of conventional war in the next twenty years leads to no simple or certain conclusions. The technical progress I have envisaged assumes that the same amount of effort that was put into the development of conventional weapons in the 1950s will be maintained, as a result of continuing international tension. One must realize that, if the arms themselves are considered as a danger, it is only because of their power to amplify political crises; it is the political crises that constitute the real danger. Until now, the twentieth century has simply allowed latent infections to develop until violent conflict becomes inevitable. Any preventive treatment would be preferable to our drastic cures.

On the other hand, I have argued that the risk of conventional war is largely dammed up by nuclear deterrence. Up till now it has played a beneficial part, in spite of—perhaps because of—the apocalyptic fears it has provoked. A peaceful future requires the preservation of the phenomenon of nuclear deterrence while eliminating the residual risks of escalation into actual nuclear war. This problem extends beyond that of conventional war, but it may be of some use here to remark that, in my opinion, its solution must come through the creation of an effective international force—a distant aim, certainly beyond 1984, but one that must sooner or later be achieved. May it be as early as possible, before a new experiment with a great world war.

Finally, I cannot emphasize too strongly that conventional war itself represents a major danger for humanity. The horror of nuclear warfare tends to make people think that a return to "good old conventional warfare" would be progress. That is to forget the hell of Verdun, Hamburg, and Dresden; events like those would be even worse in any future conventional war. As long as war remains one of the modes of international practice the essential problem consists in dissuasion from it, and the peaceful solution of international disputes. But if that aim cannot be completely achieved (and to impose peace the two adversaries must agree to it), it remains imperative to retain for conventional warfare as many as possible of the limited and humanitarian features it has often acquired in the course of history, generally at the end of great, useless, and exhausting conflicts. In this direction, and in the absence of a juridical system for settling disputes that is probably not for our century, it seems to me desirable to conserve carefully the threat of nuclear sanctions, the only ones capable of imposing on the belligerents the inhibitions needed to keep the use of force within reasonable limits, as much from the material as from the moral point of view.

THE POOR MAN'S
POWER

B Y

VLADIMIR DEDIJER
Yugoslavia

Professor Dedijer was a senior officer in Tito's guerrilla army in
Yugoslavia during World War II. A historian by profession,
he now works in Ljubljana and makes a special study of the nature of revolutions.

AMONG PEOPLE in the industrially advanced countries fears
about future wars tend to focus on nuclear weapons carried
by long-range missiles, on other possible methods of mass
destruction, and on the dangers of open conflict between
major powers such as the United States and the Soviet Union, or others. But to be concerned only with the threat
of world-wide thermonuclear war is to ignore the character
of actual warfare since 1945. Guerrilla wars have been in
progress, more or less continuously, in one part of the world
or another, but always in economically underdeveloped

ian rule but also all foreign influences and the
ctures linked to them. It also had to be an ideol-
ing the facts of life. Resistance movements sur-
n there was no discrepancy between ethical prin-
real life; otherwise they were subject to internal

ie origins of guerrilla warfare lying in the sponta-
stance and social discontents of the general pop-
i whom it depends for active support, it follows
ires of intimidation or reprisal by the opposing
ces have the opposite effect to what is intended
In Europe, when political and organizational fac-
d the guerrilla movement, reprisals by the Ger-
not to secure rule over docile masses, but, on the
o even more determined revolt than before, with
ie population joining the guerrillas. As a general
n guerrilla warfare, manpower is fundamental to
d is to be valued above control of territory. A
ar is essentially a psychological and political war.
ve role of morale in warfare is nowhere so obvi-
i guerrilla movement, which is usually confronted
enemy forces equipped with powerful weapons
; total command of the air.

uerrilla war is not a closed system, immune to
d events. Comparison of the outcome for the
nd Greek movements shows the influence of the
on, the United Kingdom, and the United States,
carried the main burden in the war against Ger-
r 1941 the Adriatic region was not regarded by
or the Germans for that matter, as a major the-
r. The guerrilla warfare sprang up before any di-
s delivered by the Allies; when aid came, it was
il importance but only secondarily significant
litary point of view. Yet the eventual success of

countries. We can see such wars continuing today and, looking at the world of the future, one can only predict a series
of guerrilla wars arising from the grave social unrest that is
latent in large regions of the earth.

It is a potent form of warfare available to the most technically backward populations and scarcely susceptible at
all to control by international treaty or other disarmament
measures. If the nuclear stalemate continues between the
major powers, guerrilla warfare will persist as the principal
military factor of our time. As we have seen in the recent
past, major powers may be drawn into antiguerrilla operations; in turn there may be intervention on behalf of the
guerrillas so that, superficially, the struggle may come to
resemble a conventional war. There is an ever-present risk
that what begins as a revolutionary civil war may grow and
grow until major powers are in open conflict. With a view
both to avoiding errors in response to guerrilla outbreaks
and, above all, to dispel conditions in which such outbreaks
are likely to occur, it is essential to understand the character
and motives of guerrilla warfare. Only on that basis, too, is
it possible to approach the question of how and where guerrilla warfare may be waged in the future.

Guerrilla Wars of the Recent Past

The distinction between guerrilla warfare and other
military and paramilitary activities that may resemble it is
best illustrated from experiences in occupied Europe during
World War II. Then, we witnessed various forms of resistance against German rule, but not all of them could be
described as guerrilla warfare. In each occupied country the
resistance movement had its own special style and waged
war in its own way. Some of these actions assumed the character of national wars continued after military defeat, with

the sole purpose of ending the German rule. Such was the case particularly in industrialized countries with good national and social cohesion (Norway, Denmark, the Netherlands, Belgium, and France). There was little or no purpose beyond the defeat of the Germans; no major change in the social patterns of these countries followed the end of German occupation. The forms of resistance in these countries could not be identified with the classical concept of guerrilla warfare, as I interpret it or as it will figure in the world of the future. To be sure, techniques used by these movements had much in common with those of guerrilla warfare: the use of irregular troops; reliance on elusiveness, knowledge of the local terrain, and sympathy from a large section of the civilian population; sabotage and ambush; capture of arms and supplies from the enemy. But both the motives and organization of these resistance activities lacked the political and revolutionary character of guerrilla warfare.

Consider now the economically underdeveloped parts of Europe during World War II. In these countries, particularly those of southern Europe, there were pre-existing social and political conflicts when the Germans or Italians occupied them. Here the resistance assumed much more clearly political forms, and in Yugoslavia, Albania, and Greece a species of guerrilla warfare appeared. Because these countries were economically backward there was little or no working class in the historical sense of that term; the guerrilla movements found their social basis in the peasantry, by far the most numerous part of the population.

Guerrilla warfare in these countries was not exclusively military in purpose, in the technical sense of the defeat of the German occupying forces. It was also aimed at the destruction of pre-existing institutions, either domestic or for-

eign in origin, that represented
tion within the countries. The
basically, a social reformer. In t
new institutions were created, p
army, molded from the guerrilla
cial and political character as w
and tactics of guerrilla warfare.

The nature of the guerrilla v
circumstance as upon ideology
Party took the leading role in
warfare there was quite differer
partisans in the occupied parts
resistance behind the German
cupied Europe, both in terms
and in the style of the milita
partisans were an auxiliary arm
directed, with the particular
the German army and its line
cert with the operations of th
cific tactical aims. By contras
slavia was self-created and had
depend. It was a new kind of
olutionary situation in occup
gans of power were formed, to

The strength of the guerr
Albania, and Greece, compa
nomically underdeveloped c
needs some explanation. S
population at large is of pri
out it, organized resistance
ous resistance is not suffic
ideology and organization di
ideology had to match the n
ulation, typically in the n

onl
soc
ogy
vive
cipl
cris

W
neo
ulati
that
regu
by tl
tors
mans
contr
more
princi
victor
guerri
The d
ous as
by str
and ha

But
other
Yugosl
Soviet
which
many.
the Alli
ater of
rect aid
of polit
from a

the movements was settled by secret agreements between the Soviet Union, the United Kingdom, and the United States, concerning the division of influence in liberated Europe. Greece was put firmly in the Western sphere of influence, and the Soviet Union refused to intervene when British troops crushed the Greek resistance movement in December 1944; the social revolution for which the guerrillas had fought did not occur. Similar motives led Moscow to send instructions to Palmiro Togliatti that the Italian partisans were not to occupy the factories of Turin and Milan, at the end of the war. Yugoslavia, on the other hand, was "fifty-fifty" in relation to the Eastern and Western spheres of influence, and no doubt this fact was of great importance in allowing the final political success of the Yugoslav guerrilla war.

Since 1945 guerrilla warfare has exhibited almost all the features I have ascribed to it in the narrow European context. It has occurred in the economically underdeveloped countries of Africa and Southeast Asia and in Cuba, all inhabited by peasant masses. It has been nationalist in spirit, directed against colonial rule or domestic governments dominated by foreign influences. The conditions for success are still the same: for example, that the guerrilla movement cannot thrive in a countryside that does not give them active support. In Malaya most of the guerrillas were Chinese and represented a minority of the local population; they were defeated. In post-1945 guerrilla movements, as in Europe, severe retaliatory measures by the opposing forces have strengthened rather than weakened the guerrillas.

Successful guerrilla warfare is essentially an art of the undogmatic and the original-minded social reformer. The political springs of guerrilla warfare have changed since 1945; Communists have not the special place they had in the European guerrilla wars. In Cuba guerrilla warfare against the former regime was started against the wishes of the Cuban Communist Party. During the Algerian war, the French Communist Party took no strong action in support of the Algerian guerrillas, even though it was backed by one-fifth of the French electorate and could have called general strikes or sought to subvert French troops in Algeria. The party's behavior was similar during the French Army's struggle with the guerrillas in Indo-China.

It would indeed be rash to generalize about the political or geopolitical connotations of guerrilla movements, or seek to apply some conspiratorial theory of history to them. Despite many similarities of conditions and governing principles, the movements in Cuba, Algeria, and Vietnam have differed markedly in their historical, social, and international traits. The relations or conflicts between guerrilla movements and great powers are, as in the case of Europe, very important but not necessarily decisive. In particular, material aid from outside is not vital for guerrillas who are accustomed to seizing their arms from the enemy and otherwise living a very simple life. (For the same reasons, any policy of bombing supposed "lines of communication" is likely to be futile.)

On the other hand, the established rule and institutions against which guerrilla warfare is directed are characteristically dependent on foreign support—political, financial, material, and often military, too. Without such support the war could not continue because if, as we have defined it, a guerrilla movement matches the political wishes of the general populations, the existing government could not survive; in other words, the event would be a brief revolution rather than a protracted guerrilla war. There are now few nations on earth in which some great power does not have an interest in maintaining the existing regime, for the principle of "spheres of influence" now extends far from the

Elbe and the Adriatic. Therefore, in many countries, one would predict that attempts at social revolution would lead to long drawn-out guerrilla wars.

The Foci of Revolution

Reformist revolutions do not occur in countries with broadly contented populations, such as the industrialized "welfare states." Standards of living of the industrially advanced countries are rising continuously. These countries have good reason to feel materially contented, though perhaps their consciences should be uneasy. With one-sixth of the world's population they dispose of three-quarters of the world's wealth, as measured by gross income. The gap is widening between the developed and underdeveloped countries.

In many underdeveloped countries, in Latin America, Southeast Asia, and parts of Africa, the social tensions associated with economic backwardness are potentially explosive. Not only is the theoretical average income very low by the standards of the industrially advanced countries, but great inequalities may exist in the distribution of that income because of the social and economic structures of the nations concerned. These structures may also be a positive impediment to increase in national wealth, because the strata of society that, by virtue of conservative patriarchal, tribal, and feudal relations, controls the surplus wealth is not prepared to invest it productively in modern ways. Just as the gap between rich and poor nations is widening, so are the richer sectors of some underdeveloped countries becoming richer while the poor grow poorer. These sources of strain are compounded with the effects of the population explosion, and the outlook for the world's poor is indeed bitter. Populations will tend to grow most rapidly in the very countries where the people are poorest; in the underdeveloped regions as a whole, the rate of growth is expected to be twice as great as in the industrialized countries.

The statistics of world poverty and exploding populations, and grim evidence at times of crisis and famine of the human realities that underlie the statistics, have long been available to the great powers. Yet they have failed to agree on a common plan for massive aid without political "strings," which would accelerate development and release hundreds of millions of people from the tightening grip of despair. Relatively little aid is available from truly "neutral" sources such as the United Nations. Aid given bilaterally, even with subjectively pure and humane motives, will tend to be seen as politically motivated by dissident elements within the receiving country—as a foreign influence of the very kind that bolsters the existing regime, the kind that nationalist guerrilla wars are fought to end. Similarly, investments of foreign capital, useful though they may be in strictly economic terms, may well become a focus for nationalist and revolutionary discontent, especially if the profits are exported and if the investors form a political alliance with the regime. Many commentators treat multilateral (United Nations) and bilateral aid as though they were merely interchangeable sources of funds for economic development, without appreciating the great subjective differences, from the viewpoint of a nationalist rebel. In short, economic aid with conscious or unconscious political connotations, just as much as military and political assistance, may reinforce rather than diminish the unrest in the poor countries of the world, even though massive economic aid is the only way of removing the long-term economic causes of the unrest.

Unless there is a fresh, concerted approach to world development, by the powerful and rich countries of East and West, neither the scale nor the manner of economic assistance is likely to prevent violent outbreaks of guerrilla-style warfare in country after country, as poverty and exploitation become worse than the peasantry will tolerate. I have already mentioned Latin America, Southeast Asia, and parts of Africa as the principal regions where such wars must be expected. Today, the population of Latin America is about 200 million; by the year 2000, it will be about 500 million, unless famine supervenes. It is simply not credible that such growth can occur without major social changes in the countries of Latin America. If those changes are not brought about voluntarily by the ruling strata, popular revolutions will occur. As it is unlikely that they will be allowed to take their course swiftly and without foreign intervention, long and terrible guerrilla wars will ensue.

Guerrilla Warfare and New Weapons

A general feature of guerrilla warfare is as plain in Vietnam in the 1960s as it was in Algeria in the 1950s or Yugoslavia in the 1940s. It is that technical superiority in arms and logistic support possessed by the forces opposing the guerrillas is quite unimportant in determining the outcome of the war. A second general feature is that military commanders tend to be blind to the true character of a guerrilla movement that they are fighting, and that they make the same political, psychological, and tactical mistakes over and over again. This is not the place to discuss guerrilla strategy and tactics in general, but something should be said about their relation to the fearsome new weapons that military science and technology are making available to regular forces.

From a military point of view the art of guerrilla warfare

is largely a matter of improvisation and opportunism. As I have mentioned, greater importance attaches to conserving manpower than to holding ground. With this principle in mind, the guerrilla is always ready, if need be, to run away and hide. Consequently, when regular ground or air forces attack with powerful modern weapons, the guerrilla makes it his business not to be present. It is possible that areas of open country which are regularly patrolled on the ground or surveyed from the air will be entirely denied to him, unless he moves in the guise of a civilian. The use of chemical agents to defoliate trees may rob the guerrilla of cover in some places. A massive sweep through a particular area by the regular forces may force him out of it, temporarily. The important point is that, unless the regular forces are prepared to lay waste to the entire country, with nuclear weapons for example (which cannot correspond with any rational political goal), there will always be somewhere for the guerrilla to conceal himself from attack by modern weapons systems. From his hiding place he can emerge at a moment of his own choosing to attack the enemy at a vulnerable point. The only strategy open to the regular forces is to attempt to eliminate the guerrillas, one by one, in infantry engagements over a huge area and a long period of time. The war becomes uncontrollably large and costly in lives and materials for the regular forces.

If the guerrilla movement (by our earlier definition) is supported by a large section of the population, it has a reservoir of manpower on which to draw. Clumsy, destructive, and impatient action by the regular forces will always tend to drive civilians into the guerrilla camps. A characteristic of advanced and powerful weapons systems by land and air seems to be that, despite the allegedly "sophisticated" electronic systems for fire control, the impression at the target is of clumsiness and purposeless destruction. For this reason advanced weapons systems may be much more of a liability than an asset in antiguerrilla warfare. And even if some highly lethal new weapon, such as a nerve gas or an infectious microorganism, were used effectively against the guerrillas, that local military victory of the regular forces would be bought at a price of moral catastrophe that would make political victory utterly impossible.

A secondary consideration about the use of advanced or novel weapons against guerrillas is that, as the guerrillas rely largely upon captured arms, these weapons, too, are quite likely to fall into their hands and be turned upon the original owners.

To sum up, guerrilla warfare is likely to remain militarily viable, whatever weapons are used to combat it, unless those weapons are so morally indefensible or indiscriminately destructive that the user forfeits all his political purposes. The well-organized guerrilla movement in a country of reasonable size, based on popular support, is indestructible except by enormously costly and protracted infantry warfare which is almost beyond the resources of even the biggest nations. In this sense, it represents the eternal truth that you cannot destroy a political belief without killing, one by one, all the people who possess it. That is something that no scientific advances in weaponry can alter, even though they make the killing easier to accomplish.

THE PERILS OF NUCLEAR PROLIFERATION

B Y

SIR JOHN COCKCROFT
United Kingdom

This chapter was completed shortly before Sir John Cockcroft's death in September 1967. As director of the Atomic Energy Research Establishment, Harwell, he played a central part in the British nuclear program after World War II. He was a Fellow of the Royal Society, a member of the Order of Merit, a Nobel prizewinner for physics, and, at his death, Master of Churchill College.

In the period from mid-1945 to mid-1967 five nations have proved nuclear weapons of different kinds: in chronological order, the United States, the Soviet Union, Britain, France, and China. Four, excluding only France, tested H-bombs during that period, and the United States and the Soviet Union also developed ballistic missiles for the delivery of these weapons. Britain is acquiring American missiles for launching from its own nuclear-powered submarines, while France is developing its own submarine missiles; the programs of both nations are modeled on the American Polaris submarine system.

There is good evidence that the nations concerned have found that, while pure fission bombs (A-bombs) can be made using either uranium-235 or plutonium, a militarily practicable fusion bomb (H-bomb) requires uranium-235 for the fission trigger. The explosive energy of an H-bomb is typically a hundred times greater than that of an A-bomb. Whether a nation has access to uranium-235 in particular is, therefore, as important a question as the availability of fissile explosives in general. Plutonium is made in nuclear reactors. Uranium-235 has to be separated from the common isotope of natural uranium, uranium-238, which is not readily fissionable. This separation requires an elaborate and very costly physical process, normally dependent, up till now, on repeated diffusion of uranium-bearing gas through porous barriers. There have been experiments and speculation about other techniques, of which only the use of gas centrifuges seems a possible alternative to diffusion. The United States, the Soviet Union, and Britain have had plentiful supplies of both weapons-grade plutonium and uranium-235 for more than a decade. France has so far developed plutonium bombs and, with the completion of the diffusion plant at Pierrelatte, built explicitly for military purposes, the testing of a French H-bomb assembly cannot be long delayed.

China, on the other hand, went straight for uranium-235 bombs; indeed, sampling of airborne material revealed that China's very first bomb (1964) was of this type. No definite information is available about the source of the Chinese uranium-235, but it must come either from a diffusion plant or from a centrifuge plant. China had therefore the potential capacity for building H-bombs and, in June 1967, it exploded its first. The Chinese are also reported to have developed short-range ballistic missiles, and Robert Mc-Namara, the former U.S. Secretary of Defense, has suggested that China will have a long-range ballistic missile within seven to eight years.

Plutonium from Power Reactors

Plutonium is an essential by-product of nuclear reactors, whatever the purpose for which the latter are built. The element is formed by the absorption of neutrons in the abundant isotope uranium-238. The plutonium production from low-power research reactors is usually negligible. With the rapid growth in electricity generation from large nuclear reactors, several countries, in addition to the five already possessing nuclear weapons, are acquiring a significant potential for plutonium production.

For each megawatt-day (24,000 kilowatt-hours) of heat output in the fuel elements of a reactor, about one gram of uranium-235 undergoes fission. This fissile uranium is partially replaced by plutonium-239 "bred" in the capture of neutrons by uranium-238. The amount of plutonium bred depends on the type of reactor. Reactors of the British "Magnox" type, which use natural uranium fuel, breed about 0.5 kilogram of plutonium each year, for each megawatt of their electrical rating, assuming that the reactors operate for eighty per cent of the year. Water-moderated reactors of the type favored in the United States, using fuel enriched in uranium-235, produce less plutonium—about 0.2 kilogram per megawatt-year. "Breeder" reactors of the types now under active development by several countries will produce substantially more plutonium per megawatt-year. A new generation of plutonium-fueled reactors is envisaged, for producing power from the abundant supplies of plutonium that are becoming available.

The potential annual production of plutonium by different countries can be estimated from data on the reactors of member nations published by the International Atomic Energy Agency; these estimates are given in the table below. The table shows the approximate potential plutonium production from power reactors in operation at the end of 1966 in countries not yet possessing nuclear weapons, and the additional potential production from power reactors known to be under construction at that time and due to start up in the period 1968–1971. From the table it appears that, by 1971, each of seven nations not already possessing nuclear weapons will have the potential to pro-

ESTIMATED PLUTONIUM IN KILOGRAMS PER YEAR

COUNTRY	From reactors in operation in 1967	From power reactors under construction
Belgium	2	—
Canada	150	500
Czechoslovakia	—	75
Germany (Federal Republic)	75	160
India	10	180
Israel	5	—
Italy	160	—
Japan	80	220
Netherlands	—	10
Pakistan	—	60
Spain	—	120
Sweden	5	120
Switzerland	—	70

No figure for the mass of plutonium needed to make a bomb has been officially released, but the French physicist Bertrand Goldschmidt has published seven kilograms as a minimum figure.

duce more than 100 kilograms of plutonium per year: Canada, the Federal Republic of Germany, India, Italy, Japan, Spain, and Sweden.

It is well known that to be useful for military purposes the plutonium must not contain too high a proportion of plutonium-240. This isotope is formed in a reactor by neutron capture in plutonium-239. In other words, if newly formed plutonium-239 is left for a long time in a reactor it will tend to be degraded, as far as military application is concerned. For the production of military-grade plutonium the fuel elements must be extracted from the reactor before their useful life for power production has expired. This inevitably requires that the uranium throughput in the reactor is increased.

One hundred million dollars will buy reactor capacity adequate for the production of plutonium for several bombs a year, and if that is linked with economic power production, the net cost of the plutonium will be arbitrarily small, dependent on the value assigned to the electricity. An assured supply of uranium not subject to international safeguards against military use is required.

The plutonium has to be separated from the uranium in the reactor fuel rods, in a chemical separation plant of some complexity. These plants exist in the United States, the Soviet Union, Britain, and France—and also at Trombay in India and at Mol in Belgium, the latter being the Eurochemic plant, a joint project of thirteen European countries. The Indian plant is reported to have cost about $8 million and the Eurochemic plant about $36 million.

The Price of Nuclear Forces

The possession of military-grade plutonium is, however, only a first step in the bomb-making process. The plutonium has to be assembled with an isotopic or other source of neutrons to initiate the chain reaction, and surrounded with high-explosive charges fired by very sophisticated electronic equipment and with built-in safeguards to prevent accidental explosions. The whole assembly will then have to be tested, and if the country concerned has adhered to the partial test-ban treaty, this can only be carried out underground, at not negligible additional expense. All of this is very costly, judging by the experience of the existing "nuclear powers," and the minimum expenditure, including a substantial outlay on a nuclear weapons establishment, might well be at least $300 million. Even greater sums are involved if uranium-235 is also to be manufactured and H-bombs are to be assembled and tested. The cost of the French diffusion plant was of the order of $1 billion.

Nuclear bombs are, of course, of no use without means of delivery. The British V-bombers built for this purpose are said to have cost about $1,400,000 and are becoming obsolescent. Britain is therefore having to build a small fleet of Polaris submarines, at a comparable cost. The French are going through the same sequence of building obsolescent bombers and new missile-carrying submarines, at a similar level of prices. Nations acquiring nuclear bombs for the first time would probably follow the example of France and use bombers, in spite of the high degree of vulnerability to antiaircraft weapons.

The cost of a nuclear weapons program is primarily in the capital cost of providing the facilities for their development, production, and delivery: the cost of the weapons establishment, the cost of the plutonium separation plant, and the cost of the bombers or other delivery systems. In addition to these there will be the operating cost of the establishments for research and weapons development; the cost of the proving tests; the operating cost of the air force or rocket force. Compared with all these additional costs the cost of making plutonium itself is rather trivial, especially if it is linked with the production of economically useful electricity. The Institute of Strategic Studies, London, has estimated that the cost to Britain of the nuclear strategic forces was of the order of £100 million ($280 million) per year, in 1966–1967.

Any nation aiming at more than a token show of nuclear strength probably has to be prepared to spend at a comparable rate. It might be possible, at much less cost, to stage the test of a primitive bomb, in an attempt to create an illusion of nuclear armament, but to try to bluff about nuclear weapons while lacking a militarily significant delivery system could be an extremely dangerous policy.

Some indication of the cost of producing additional nuclear weapons once they have been developed is given by the U.S. Atomic Energy Commission's estimate of the cost of nuclear explosives that could be provided for a "Plowshare" program—that is, in the suggested use of nuclear weapons for peaceful engineering works. Figures are given by Dr. Inglis on page 47.

There have been suggestions about possible "cut-price" nuclear weapons systems, and in particular about radiological weapons. Nuclear bombs inevitably produce radioactive fission products, and bombs of more than one megaton could produce lethal radioactive contamination over an area of hundreds of square kilometers. By analogy, radiological weapons would carry not nuclear explosive but concentrated radioactive material for dispersal over a target area, to render it uninhabitable for a short or long period. The only comparatively cheap source of highly radioactive material is the waste fission products from a nuclear reactor.

There would seem to be no military advantage in dispersing radioactive fission products derived from such a source rather than by dropping nuclear bombs. The problems of shielding the radiation of the fission products in aircraft, or in rockets before launching, would be formidaable. They can therefore be neglected as potential weapons.

Proliferation and Nonproliferation

Unfortunately there are many countries in the world that could afford to develop nuclear weapons and minimal delivery systems, particularly if they became convinced that national security depended upon doing so and if they believed that they could economize in other military expenditure. Although competent scientific manpower is required and the relevant engineering is not trivial (especially for the production of uranium-235), there are no "secrets" inaccessible to an industrial nation prepared to carry out the necessary research and development.

To speculate about which nations might seek to acquire nuclear weapons during the next few decades is somewhat invidious and must depend on many political and economic assumptions. Alignments of nations in alliances and in conflicts plainly exert restraining or provocative influences. From a global viewpoint, the chief cause for anxiety is the likelihood of a chain reaction in which the acquisition of nuclear weapons by one new country would provoke other nations to follow suit. For example, if three nations made nuclear weapons for the first time in the 1970s, ten might do so in the 1980s and thirty in the 1990s.

The chance of nuclear war breaking out, and possibly engulfing a large part of the world, must inevitably increase with the number of nations possessing nuclear weapons. The risk is aggravated by the suspicion that smaller countries acquiring nuclear weapons would be unlikely to develop the sophisticated command and control systems of the kind possessed by the United States, the Soviet Union, and Britain. These systems are strongly biased against the possible delivery and detonation of the bombs as a result of false information, unauthorized action, or an engineering fault. The lack of such systems would greatly increase the risk of nuclear war by accident. Another important factor is the likelihood that some nations acquiring nuclear weapons would have a less responsible form of government, or be susceptible to revolutions that might bring reckless men to power. In that connection the use of nuclear weapons in civil war cannot be entirely excluded.

We have several examples of powers that are mutually hostile at present, for example Israel and Egypt and, to a lesser extent, Pakistan and India. The tensions that have been generated in the past would be enormously increased if they had the possibility of dropping only a few nuclear bombs on one another.

There are also strong economic arguments against the development of nuclear weapons by underdeveloped countries while their standards of living are so low. They need the whole of their scientific and technological effort for the development of their economy rather than devoting a large proportion to nuclear weapons programs.

A further argument is the complete uselessness of nuclear weapons for any purpose other than mutual destruction. They would have been of little effect if India had possessed them and used them to try to stop the Chinese incursion over their border.

At the beginning of 1967 the risks and costs of the spread of nuclear weapons were sufficiently plain, the main nuclear powers in sufficient agreement, and the military ambitions of potential bomb-making countries sufficiently restrained, to make possible serious international negotiation of a nonproliferation treaty. The objections raised at the time, and the answers to them, are relevant to the successful conclusion and long-term durability of such a treaty.

The West Germans, in particular, have raised two objections. The first is that the development of civilian nuclear power programs would be handicapped by a nonproliferation treaty. The facts are, however, that the installation of nuclear power stations is proceeding rapidly in Italy, Japan, Germany itself, India, and Pakistan by purely commercial arrangements with American, British, and Canadian industrial organizations, while indigenous development programs are making impressive progress in several countries. The overlap in "know-how" between weapon building and the development of efficient power-producing reactor systems is nowadays virtually nonexistent.

The second German argument concerned inspection. Safeguards against the diversion of plutonium for military purposes will be applied by the International Atomic Energy Agency (I.A.E.A.) in Japan and India, and by Euratom (the nuclear wing of the European Common Market) in the Common Market countries. The reactors are, in such cases, subject to inspection, without warning, if need be, so that any abnormal rate of fuel turnover, or other indicator of suspicious behavior, will come to light. The countries concerned could apply their plutonium to military use only by breaking their safeguards agreement and in the full knowledge of the rest of the world. The I.A.E.A., which includes all countries of East and West with serious nuclear interests, excepting only China, is the natural organ for policing a world-wide nonproliferation treaty. Indeed, it was set up in 1956 for the very purpose of making the peaceful benefits of nuclear energy available to all comers, while insuring that assistance and fuel received would not be misapplied to military purposes.

The Germans have objected that inspection by foreign experts under the safeguards system of the I.A.E.A. would give away commercial secrets in the power stations. This argument does not hold water. The details of nuclear power reactors are usually published in the technical literature as soon as they come into commission. It is only in the design and building phases that commercial secrecy is important and, at that stage, as no fuel is being burned, the question of inspection does not arise. Even if a nation should have good reason for wishing to retain commercial secrets after a reactor has been commissioned, it is fully entitled, under the I.A.E.A. arrangements, to object to individual inspectors of whose integrity or commercial connections it is suspicious.

A third argument is that nuclear explosives will become important for civil engineering purposes—to blast out canals, harbors, or storage vessels, or to break up oil-bearing rocks to release the oil. It is claimed that countries not possessing nuclear explosives would be at a considerable disadvantage in working in these fields. These hypothetical applications have been much publicized in the American Plowshare program. It has not yet been found possible, however, to produce nuclear explosions without releasing large amounts of radioactive fission products, though the amount released per unit of explosive power is no doubt being reduced by research. It seems very unlikely that agreement would be obtained to the use of such nuclear explosives in populated regions unless radioactive contamination could be kept to a very low level. Their use to break up rocks underground would give rise to radioactive contamination of the oil, or whatever the product was, and this would be equally objectionable to the users.

Assuming that all these difficulties could in time be overcome, it should be possible for nuclear explosives to be provided by those in possession of the technology, to the I.A.E.A., for sale to those who wish to carry out such civil engineering applications.

Responsibilities of Existing "Nuclear Powers"

A general political objection to the nonproliferation treaty is that the existing "nuclear powers" should not expect the countries that do not have nuclear weapons to be the sole contributors to the reduction of the risk of nuclear warfare. The unavoidable implication that some countries are "entitled," for historical reasons, to possess nuclear weapons is hard enough for other countries to accept. It is said, therefore, that the "nuclear powers" should themselves make a contribution by reducing stockpiles of fissile materials and nuclear weapons, especially since they already possess such an enormous capacity for "overkilling." There is certainly a great deal to be said for this argument.

There is a serious danger that the trend will be the other way. The development of antiballistic missile systems in the United States and the Soviet Union would lead to fresh competition in nuclear armaments. Not only would such defensive systems themselves employ nuclear warheads, but the obvious countermeasures would include the deployment of many more offensive ballistic missiles carrying nuclear warheads, with a view to saturating the defences (see the following chapter). The further great increase in the stocks of nuclear weapons would constitute an enormously greater potential danger to the whole world, should nuclear warfare develop. The risks of proliferation are therefore not exclusively those of the acquisition of weapons by further countries, even though that is what the term is usually taken to mean. The "nuclear powers" have a great responsibility to curtail and to reverse the proliferation of their own weapons.

THE OUTLOOK FOR NUCLEAR EXPLOSIVES

B Y

DAVID INGLIS
United States

Dr. Inglis is a senior physicist at the Argonne National Laboratory, Chicago. He is a member of the editorial board of the Bulletin of the Atomic Scientists and a former chairman of the Federation of American Scientists.

A SOUND JUDGMENT of the likely future of nuclear weapons must be based on an appreciation of the basic processes involved in present nuclear weapons. The existence of bulk matter, of chemical reactions, of life, of a universe sparsely studded with enormous furnaces known as stars, and of nuclear weapons, depends on the availability of certain kinds of particles pushing or pulling on one another with particular kinds of forces. The particles and forces are limited in kind.

The atom has its encircling electrons bound to its heavy

central nucleus by electric forces. Readjustments of the energy of this electric attraction provide the power of the oil burned in a stove, or of a chemical explosive such as dynamite or TNT. Such burning gives off a relatively modest amount of energy per atom involved, perhaps one or two "electron-volts" for each atom. The central nucleus is held together by forces so much stronger than electric forces that, when rearrangements occur among the protons and neutrons of which the nucleus is composed, the energies involved are measured in millions of electron-volts.

The forces of nature seem to make possible the building of only about a hundred kinds of chemical elements—enough to make chemical and life processes sufficiently varied and still not too confusing. The trick is a very interesting one. There are two quite different kinds of forces between the particles within the nucleus. These neutrons and protons attract each other with very strong "nuclear" forces, but the protons push each other apart by electric forces. The cohesive nuclear forces dominate in very small nuclei but increase more slowly than the disruptive electric forces as nuclei grow in size. Nuclei with more than about one hundred protons (and a slightly greater number of neutrons) do not exist, because if they were formed, the electric force would quickly split them apart.

The result of these immutable laws of nature is that, when dealing with the light nuclei containing only a few particles, we can gain useful energy from the forces of attraction by *fusion* of nuclei to make a bigger nucleus. At the other end of the list of nuclei, among the very heavy ones containing many protons and neutrons, we can gain useful energy if we can disturb a nucleus in such a way as to cause the electric repulsion to cause *fission*, to "split" the nucleus into smaller nuclei. Among the heavy elements, uranium-235 and plutonium-239 have nuclei that, when "prodded" by neutrons, will undergo fission. In the process, the fragments of a nucleus release a couple of other neutrons, which then hit other nuclei and cause further splitting —the famous chain reaction.

In between the very light and the very heavy nuclei there is a long middle part of the list where the cohesive and disruptive tendencies just about cancel one another; the nuclei are very stable and nothing can happen that involves large amounts of energy. The entire list of available nuclei is well known, and if we except radioactive decay, which is of limited usefulness, fusion and fission are the only ways of obtaining such large amounts of nuclear energy from the materials and forces that can be manipulated on earth.

Existing Nuclear Weapons

Fission came first, in the original A-bomb. Whether an explosive chain reaction will work in a mass of uranium-235 or plutonium-239 depends on several numbers characterizing the behavior of nuclei. Many of the Western scientists who devised the first A-bombs during World War II rather hoped that nature had determined these numbers in such a way as to make the bomb impossible. It turned out otherwise, and the very first bombs were a thousand times as powerful as the biggest chemical-explosive bomb that had ever been made, the twenty-ton "blockbuster." The power was equivalent to that of 20,000 tons of TNT— twenty kilotons, as one says. Such cold numbers mean little until translated into terms of the depth of human tragedy caused at Hiroshima and Nagasaki.

The fusion process is the basis of the much more powerful H-bomb, typically of megaton force—equivalent to a million tons or more of TNT. It is achieved not by neutrons but by banging nuclei of light atoms together so violently that they fuse. A very energetic impact is required to overcome the electric force between two nuclei that tends to keep them apart. The atoms have to be set in very violent motion; in other words the material has to be raised to a very high temperature. That is why the fusion reaction is often called a "thermonuclear" reaction. In practice, a uranium-235 A-bomb serves as the "trigger" to create the very high temperature and detonate the fusible material of the H-bomb.

The greatest practical difference between the A-bomb and the H-bomb is that the power of the A-bomb is limited by its having what is known as a critical mass, while for the H-bomb there is no such limit. An arbitrarily large mass of the appropriate light elements of an H-bomb can be assembled without danger of premature detonation. A large mass of plutonium, on the other hand, will explode spontaneously. The critical mass at which the chain reaction will begin depends on the supply of neutrons and their fate. A neutron born in the middle of a sphere of plutonium may or may not hit a plutonium nucleus, depending on how big the sphere is. It is like a bullet fired in a forest: it may miss the trees and escape into a field beyond unless the forest is quite large. For plutonium metal of normal density and with certain materials around it, there is a particular size that will let so few neutrons escape that the chain reaction will start. Above this size, the nuclear explosion would occur spontaneously, for there are always a few neutrons around to get it started. For a slightly smaller size, nothing drastic happens. But if the smaller size is compressed (like having the trees in the forest closer together), it may become supercritical and intercept enough neutrons to explode. The sudden compression or sudden assembly of separate pieces is accomplished by the use of chemical explosives.

There is also the economic difference that the main bulk of the light elements used in an H-bomb is much cheaper than the specially prepared uranium or plutonium used in an A-bomb. As Americans put it, the H-bomb gives "more bang for a buck." It appears as the "big economy pack" in the price list for nuclear explosives that may be made available by the U.S. Atomic Energy Commission for large-scale underground "nuclear engineering." The prices (which follow a straight line on a logarithmic graph) are typically:

KILOTONS	PRICE
10	$350,000
200	500,000
2000	600,000

The last big increase thus costs relatively little. The starting price is presumably the approximate cost of the A-bomb trigger, after the cost of getting into production has been written off.

A-bombs and H-bombs also differ in the proportion of radioactivity they produce. From a bomb burst high in the air, the chief radioactivity is that of the fission products, the radioactive fragments into which the uranium or plutonium nuclei split, in the A-bomb or in the fission trigger of the H-bomb. The H-bomb may throw much of its fission garbage higher into the upper stratosphere, where it may remain long enough to lose much of its radioactivity. The most potent fission products decay in a few hours or days, but some—strontium-90, for example—continue little diminished in power for many years. The fusion reaction in the H-bomb produces mainly the light radioactive substance tritium, a gas (a form of hydrogen) that mixes in the atmosphere and does not settle back to the ground. The abundant neutrons released in the fusion reaction also react with nitrogen in the air to produce carbon-14, with long-lived radioactivity.

A "clean" bomb is one that produces considerably less than the normal amount of radioactivity in proportion to its explosive power. If we forget the less important tritium, a large H-bomb may be said to be relatively "clean" if it releases only the fission products of its A-bomb trigger— perhaps less than the radioactivity produced by the Hiroshima bomb. But we must still be wary of overconfidence in the word "clean"; it is a relative term.

"Dirty" on the other hand, may mean very dirty indeed. Beyond the simple A-bomb and the less simple H-bomb there is a further combination that may be called the fission-fusion-fission bomb, or the jacketed H-bomb. This is a cheap and dirty big bomb that takes advantage of the fact that, although the common form of uranium-238 does not undergo fission with the relatively slow neutrons available in an A-bomb, it will do so in response to the more energetic neutrons made in the fusion process in an H-bomb. Thus this cheap material may be wrapped around an H-bomb to use up the escaping neutrons. It adds enormously both to the power of the explosion and to the radioactive contamination. In addition to its cheapness, the jacketed H-bomb appears to have other military advantages, for it was adopted and probably figures importantly in arsenals. Its main advantage may be compactness, arising from the fact that fission creates much more energy per atom than fusion does, even though a uranium atom is only about the same size as the light atom lithium involved in fusion. Another consideration is that a jacket, or "tamper," of heavy material is needed in any case to prevent the insides from blowing themselves apart before the chain reaction has had time to develop, and the heavy jacket may as well be something that makes still more energy.

The pace of development of nuclear weapons was impressive in the early years. Scientists had first begun to understand the nucleus properly when the neutron was discovered in 1932. Seven years later the fission of uranium was discovered, and by 1942 the first slow, nonexplosive chain reaction of uranium fission had been achieved. Three years later came the A-bomb; after seven more, the H-bomb.

Since 1952, there has been nothing really new in explosives. One might ask, have the military scientists been dawdling? The answer is no. Fission and fusion have already been used, and there isn't anything else. There have been refinements, of course, mostly directed toward making nuclear weapons of a given power more compact and less radioactive. The internal combustion engine was introduced to drive automobiles before the turn of the century and the electric self-starter around 1920; since then there has been nothing really new in automobiles, only refinements. It is much the same with nuclear weapons.

The cold facts concerning the destructive power of present bombs are well known, but too frequently forgotten. Although the range of nuclear weapons that have been tested extends from under one kiloton to fifty-seven megatons, the most important part of the range for distant strategic attack extends from one-third to ten megatons. The early models of American submarine-based missiles carry one-third-megaton warheads, later models nearer one megaton, with enough accuracy for anticity attack. Most American land-based intercontinental missiles carry about three megatons with greater accuracy so that they can be used against missile sites. Some of the largest missiles, including Soviet missiles, which are fewer and heavier than American missiles, carry more, and some long-range bombers, which are not yet obsolete, carry two ten-megaton bombs.

A one-megaton bomb can cause severe blast damage, sufficient to knock down brick apartment houses, to a distance of five kilometers over an area of seventy square kilometers. The radius for such damage for a ten-megaton explosion is about eleven kilometers, and the area 400 square kilometers—six times as great. This illustrates a "law of diminishing returns" for the size of bombs: ten times the power obliterates only six times the area. The wasted energy of the larger bomb goes into pulverizing the central part beyond complete destruction, though it may be "useful" against underground missile sites, or for caving in blast shelters if these ever became part of the civil defence program.

Blast damage is not the only, and perhaps not even the most important, kind of damage from a large nuclear burst. The radiation from the bomb not only causes radiation sickness and horrible skin burns such as were experienced at Hiroshima and Nagasaki but it also causes fires to break out at quite large distances. The distance to which the sudden burst of heat radiation sets fire to buildings and trees depends on the weather and the dryness of the foliage, but it is something like three times the radius of severe blast damage, or nine times the area, and thus covers an area of well over 3000 square kilometers for a ten-megaton bomb.

No fire department can cope with thousands of fires at once, and in such a situation they spread madly. In some of the incendiary-bomb raids in World War II there developed "fire storms" engulfing large areas. In a fire storm the combined heat of many individual fires makes a strong wind toward the center of the burning city, which fans the flames. When many fires are started at once by the single flash of an H-bomb, the number of casualties per square kilometer by burning or suffocation in the resulting fire storm would probably be almost as high as in the central blast area, for only from the edge of the fire storm would people be able to struggle to safety. Because of the fire storm, one ten-megaton bomb (or about six one-megaton bombs) should suffice to destroy a megalopolis of 3000 square kilometers. There are already so many bombs in the arsenals of the nuclear giants that several times this destructive power can be targeted on each megalopolis, with plenty left over for destroying the smaller cities several times over, too.

Future Nuclear Weapons

So much for the general principles available and their implementation in existing bombs. Nuclear weapons of the future will continue to be based on fission or fusion as the source of their energy. It is true that other ways of converting matter into energy are known to physicists, the most spectacular being the mutual annihilation of matter and "antimatter." Particles called antiprotons, antineutrons, and antielectrons (positrons) can be made in minute quantities and at great expense in high-energy experimental machines, and when one of these encounters its corresponding normal particle they both disappear, releasing a burst of energy. But the engineering problems of making and storing antimatter in militarily interesting quantities are so absurdly great that the possibility can be safely discounted for the foreseeable future.

It is likely, too, that existing explosives—uranium and plutonium for fission and lithium and heavy hydrogen for fusion—will continue to dominate in weapons. It is in principle possible that man-made elements heavier than plutonium might be used with the slight advantage of hav-

ing smaller critical mass, but this is not a practical possibility, because it need not mean a much smaller weapon when
the detonating device is included, because those heavier
elements decay rapidly and thus cannot be stockpiled,
and because their production and isolation in more than
microscopic quantities is prohibitively expensive. For these
reasons the postulated "californium bullet," for example, is
no more than confusing fantasy.

Within the limitation of fission and fusion of the usual
materials as the only energy sources, variations can be
sought that may or may not have military significance.
These include greater efficiency in using the nuclear explosives, and the development of very large or very "dirty"
bombs, and very compact or relatively "clean" bombs.

The efficiency of existing bombs falls far short of one
hundred per cent because the bomb blows itself apart so
quickly that many nuclei do not have time to react. Much
of the past effort to improve the bomb has been in this direction. The nature of recent testing suggests that the efficiency factor has been pushed about as far as is practicable;
the situation is similar to that in automobiles—the current
models are new in style but it is hard to discern any real improvement in terms of mileage per gallon.

For some purposes there will be a tendency toward more
fusion and less fission, but for the biggest explosions the
"dirty" bomb with much fission will probably be with us
for a long time. There is no limit to how powerful an H-
bomb may be made, nor to how dirty a "dirty" bomb may
be made. It is possible that the arsenals of the future will
include warheads much more powerful than present
bombs. To put this possibility in perspective, one must,
however, appreciate that a single large bomb is powerful
enough practically to obliterate a large metropolitan area.
There is thus no incentive to provide more powerful war-

heads than those in service, for the sake of destroying cities, or, more precisely, for being prepared to destroy them as part of the nuclear deterrent. But it is conceivable that incentives could be found for destroying still larger areas, such as whole forest regions, so as to ruin a land and deny sustenance to survivors of an all-city attack. During the last period of atmospheric testing the Soviet leaders talked of a one-hundred-megaton bomb, but actually did not go quite that far (fifty-seven megatons was the biggest they tested). The bigger bomb would have been feasible on either side, had this sort of incentive been felt. Such a bomb in appropriate weather would kindle forest fires over a radius of about 150 kilometers. It might also be useful in crushing blast shelters. If such a bomb is not made in the future, this will probably be either because restraining steps have been taken toward disarmament or because of the "law of diminishing returns"; six ten-megaton bombs would ignite as great an area as one dirty one-hundred-megaton bomb, with less world-wide radioactive fallout to react on the attacker.

It is technically possible to go very much further still, in deliberately making a very dirty bomb with the purpose of radioactive destruction of life on an entire hemisphere (northern or southern). In particular, the element cobalt when activated by a burst of neutrons from an H-bomb becomes a singularly powerful radioactive emitter of gamma radiation. The cobalt jacket would be wrapped around a big H-bomb, and there is no definite limit to the possible size of such a "cobalt bomb." Nevil Shute's novel On the Beach, set in Australia, carried a powerful message concerning the threat of nuclear war, but with a gentle euthanasia substituted for the horror of more likely forms of nuclear war and with some artistic license in the way the radioactivity destroyed all life on earth. The two main inaccura-

cies are that the attacking nation would have had no incentive to enter into a suicide pact with its enemies and
that the nature of the atmospheric circulation would have
confined the destruction to the northern hemisphere, at
least for a longer time. Unless we can imagine an extremely
hostile situation between the northern hemisphere as a
whole and the southern hemisphere as a whole, we have
little reason to worry that any nation will prepare a hemisphere bomb of this sort.

The strange name "doomsday machine" was introduced
into nuclear strategy discussions to refer to a hypothetical
instrument to obliterate life on the earth at the push of a
single button, a sort of *reductio ad absurdum* of the nuclear
arms race. Such an infernal machine is not technically absurd. There is little doubt that it could be constructed—
for example by means of a few widely dispersed cobalt
bombs. It is of no use to anybody not bent on suicide, unless it be for blackmail by someone who can put on a con-
vincing act of courting suicide, and it is not a serious candidate for future arsenals.

Improvements in the triggering process for H-bombs are
difficult to predict with confidence, though it seems very
likely that the possibilities have been practically exhausted.
To be sure about sources of energy we only need to take
inventory of the storehouse that nature has provided, but
to anticipate triggering processes we have to predict limitations on furure cleverness, and that is more difficult. The
brute-force method of using chemical explosives alone—
with a view to eliminating the fission stage of the present H-bomb—has not worked and is not expected to. I
have already explained the need for violent collision to
achieve fusion between nuclei. When the nuclei are as far
apart as the size of an atom, the energy of the electric repulsion between them is only a little larger than typical

chemical energies that also arise from electric forces at atomic distances—a few electron-volts. But nuclei are ten thousand times smaller than atoms, and the energies required to bang them together are more than ten thousand times the energy per atom given by a chemical explosion. For this reason a source of energy greatly surpassing chemical sources is needed to create the very high temperature to start the thermonuclear reaction between the light nuclei in an H-bomb. An A-bomb provides a concentration of energy about a million times as great as is available in an ordinary chemical explosive. There exist tricks for concentrating the energy of a large chemical explosive in a rather small volume, but it seems very unlikely that a factor of many thousands or a million can be gained in this way.

Beyond this one can imagine electromagnetic methods for concentrating energy, some of which have been tried in laboratory research on controlled, nonexplosive thermonuclear reactions. This field of technical development is being enthusiastically pursued on the basis of healthy, world-wide cooperation, in spite of the formidable difficulties of the instability that arises when one tries to concentrate sufficient energy in a small space to make a fusion process start on a modest scale in low-pressure gas. Although the situation is very different inside a solid body, it seems likely that difficulties of this general nature will also frustrate attempts to detonate an H-bomb without a fission trigger by electromagnetic means. On this point we cannot be sure, but even if such attempts should be successful in a large experimental device, it seems unlikely that the power plant to produce the electromagnetic heating could be miniaturized enough to be put in a practicable bomb or warhead. If both of these unlikely possibilities should materialize, we would have a somewhat cleaner and perhaps cheaper warhead—nothing to revolutionize warfare again.

The fact that it would require no fissionable material would make the manufacture of H-bombs possible in countries that do not have the appropriate fissionable materials available. Less improbable, perhaps, is the eventual design of efficient H-bombs using a plutonium trigger if, indeed, present H-bomb triggers are essentially uranium-235, as is suggested by Sir John Cockcroft in the previous chapter. Then, with the rapid development of power reactors in many countries, which are potential sources of military-grade plutonium, it would become much easier for those countries to acquire H-bombs.

Thus an A-bomb has been necessary to trigger an H-bomb up till now, and will probably remain the only way to do it in the foreseeable future or, if not the only way, the only compact and convenient way. It is important to make this clear because the claim has been made that a fission-free fusion bomb might soon be developed. The claim was made most loudly by Senator Dodd of Connecticut while opposing the comprehensive test ban treaty in 1959. He implied that, if underground tests were allowed to go on another three years, such a bomb could be developed. They have and it hasn't, and for the reasons stated it is not expected.

In the category of "tactical" nuclear weapons, intended to be used as firepower for ground armies, some fantastic claims have been made about new possibilities, particularly concerning the "neutron bomb" that might slowly kill soldiers by neutron irradiation without doing as much property damage as is normal with nuclear weapons. Here the dream of a really cheap fission-free fusion bomb has been hailed as making possible greatly increased firepower on the battlefield. Aside from the technical implausibility, this enthusiasm seems to be based on a lack of appreciation of the destructive power of even "small" nuclear weapons. They may be "only" about as powerful as the Hiroshima bomb and on down to a tenth or even a hundredth of that power. They have been made compact enough to be shot from large cannon (or presumably to be carried on the backs of saboteurs), but they will probably not be made much smaller. If ever used in more than extremely modest numbers they will make battling armies obsolete and defence of territory equivalent to its devastation. Field commanders are subject to the temptation to use nuclear blasts for special purposes such as creating transportation bottlenecks, but if once they were used in war conditions, escalation would be almost automatic up to the big nuclear weapons that hold the greater interest and terror. To avoid this, if there must be conventional wars at all, it is vitally important that they be restrained to stay below the well-defined line between conventional and nuclear weapons.

About the likely course of the future development of new kinds of nuclear bombs and warheads, as distinct from delivery vehicles, we can say in summary, then, that there seems to be no room for great new surprises, if we shun technical and strategic absurdities. Nuclear weapons of a given power may become slightly more compact, but not much more. The range of power per bomb can be extended, but it cannot be very much extended usefully because present bombs are already so terribly powerful. It is not impossible, though it is unlikely, that the requirement for a fission trigger of an H-bomb can be eliminated to make H-bombs cheaper, but they are already remarkably cheap relative to their destructive power, so that the increased cheapness, if it should materialize, would be important only in hastening proliferation of nuclear weapons among countries not yet possessing them.

Must the Bombs Multiply?

One might be tempted to conclude from this that there is nothing to worry about, that the worst of nuclear weapons have probably been developed, and that we've become accustomed to living with them. The conclusion is far from justified. The world situation that we have been so lucky to live through these last few years is an awfully unsafe one in which no sane race of beings would choose to live if it could help it. Our having been lucky does not mean that we are safe, even though most of us have permitted ourselves to become unaware of the nuclear threat under which we live.

There is already enough explosive power in the arsenals of the nuclear giants to provide the equivalent of over a hundred tons of TNT for each inhabitant of the world. An ounce of TNT exploded close to a person can be lethal. The total is already well over a million megatons and may soon, for reasons we shall see, be approaching a hundred million megatons. Less than ten years ago we were talking of stockpiles of ten thousand megatons as potentially almost unimaginably destructive, with one megaton being capable of destroying a large city.

Up to now, most of the missiles of each of the nuclear giants, the United States and the Soviet Union, have been aimed at the other giant, and the main danger, that of an outbreak of nuclear war between them, has been deterred by the respect each has for the power of the other. This is the apparently stable balance in which we have developed too much confidence. The prospect has been that a war would bring damage almost entirely to the giants and perhaps some of their immediate allies, with some serious local fallout spilling over the borders and with long-range fallout remaining not really very serious for most other countries. Here the most serious threat to the distant countries would be economic dislocation, unless it be some unanticipated world-wide plague. Bad as the prospect has been, it has been one that many of the countries of the world could reasonably view with greater complacency than could the nuclear giants who would bear the brunt of the attacks.

But in the absence of arms control, the moderately good side of even this bad situation cannot last. Even though the kinds of nuclear explosives may change very little, the increasing number of weapons can make the situation much worse in at least two ways, each of which makes the other more serious. The first is that the weight of explosives used in a prospective war between the nuclear giants will become so enormous that there will be "no place to hide" —that all countries will be disastrously damaged by the fallout. The second is that there will be so many nations equipped with nuclear weapons that the outbreak of nuclear war between some pair of them, possibly triggering one between the nuclear giants, will become very likely. Sir John Cockcroft deals with this proliferation in the previous chapter.

As matters have recently stood, there appeared to be signs that the total power of the nuclear weapons in the arsenals of the nuclear giants was leveling off, as though each side was becoming content with its possibility of inflicting terrible damage on the other and found it unnecessary to expend resources on a further build-up if the other side did not. The fact that the two opposing arsenals were not equal appeared to be considered relatively unimportant because the smaller was so enormous in its destructive power as to provide adequate deterrence.

Recently a new element has been introduced into the balance—the possibility of providing a partial defence against oncoming ballistic missiles. Although the idea of self-defence sounds innocuous, and an impenetrable defence could provide an entirely different basis for world stability, a merely partially effective defence can curiously be a very upsetting element in this otherwise fairly stable balance. There is no prospect of a completely impervious defence, but the possibility of a partly effective defensive system, based on the "antiballistic missile," or ABM, is at hand for both of the nuclear giants. Both the Soviet Union and, more recently, the United States, have started ABM deployment on a relatively small scale, probably for political more than military reasons. It is to be hoped that the scale will remain small, for a serious offence-defence race—as a new dimension of the arms race—would drastically increase international tension and the nuclear threat.

If, starting with stabilized numbers of intercontinental missiles, one side installs a substantial ABM system, the second side can respond by building more missiles to penetrate the defence and the first side is no safer than before. The interception by ABMs is estimated to be sufficiently inefficient for the cost to the second side of building the additional intercontinental missiles to be less than the cost of the ABM system to the first side. On a cost-effectiveness basis, there is thus no rational reason for the first side to start this step in the competition in the first place. However, military demands are seldom long delayed by cost-effectiveness arguments, and the usual experience is that what can be built will be built. The hope of arms control is to break this trend at some point, but, lacking such artificial restraint, the likelihood is that the initiative in starting ABM deployment will be answered both by more intercontinental missiles and by some ABMs on the second side, to which the first side will respond with more ABMs to try to stop the missiles and more missiles to get past the ABMs, and so forth. Thus, in place of an arms race practically stopped, the ABM can, and probably will, carry a new arms race to entirely new dimensions.

The important point is that this process adds enormously to the numbers of nuclear weapons that would be used in a war between the nuclear giants and implies an amount of fallout that would spare no nation on earth, or at least none in the northern hemisphere.

The ABM possibility is thus a destabilizing factor, an effect arising partly from the uncertainty of performance of the ABMs. The problems of delivering and intercepting nuclear warheads are discussed at greater length in the next chapter. Suffice it to say here that future developments will probably involve several stages of countermeasures, the anti-anti-anti-ballistic missile, or methods to foil penetration aids and then ways to foil these methods, and that the ABMs themselves will have nuclear warheads.

For the types of ABMs that explode in the atmosphere and fairly near the targets they are trying to protect, the explosive power will be limited by the need to avoid damage by the ABM itself to the city being protected. Such close-in ABM bursts are being planned because of the necessity of distinguishing between an intercontinental-missile warhead and its accompanying swarm of light-metal "decoys" by the way they slow down in the atmosphere. Being limited in power, such an ABM must come fairly close to the warhead to destroy it, perhaps within a few hundred meters or so (the size of the fireball). But there will also be ABMs that explode outside the atmosphere far from the defended area. They can be made very powerful, and the X-rays that they emit can be intense enough to damage an oncoming intercontinental missile at a distance of perhaps several miles. The desire to put up such an instantaneous X-ray shield to intercept as much of the swarm of unidentified decoys, including somewhere the warhead itself, as may be aimed at a large metropolitan area provides the incentive for using powerful warheads in the ABMS. Their power will be limited by the instantaneous radiation damage on the ground and people below, from the gamma rays (that make skin burns and leukemia as at Hiroshima), and from heat that starts fires. At heights of hundreds of kilometers, even these limitations permit bursts of tens of megatons, perhaps more powerful than the oncoming intercontinental missiles against which they are deployed. Even though they are exploded above the atmosphere, about half of their radioactive products descend into the atmosphere and contribute to fallout.

Such an ABM "shield" is only an instantaneous flash, so at least one ABM is needed for each oncoming intercontinental missile (since they won't attack in squad formation). An intercontinental attack may be concentrated on a region, so the defence might desire as many of these ABMS in each region to be defended as there are intercontinental missiles in the opponent's arsenal. Thus the potential military demands become enormous indeed in this defence-offence race. It seems likely that the cheapness and compactness of a "dirty" bomb will lead to its use in the above-the-atmosphere ABMS. It is conceivable that one day the size of the world's arsenal will be limited only by the available amount of fissionable material, although the limitation is more likely to remain, for the time being, in the cost of the delivery vehicles.

Without an ABM race, the hope remains that such enormous quantities will never be prepared for delivery, in keeping with the tendency for numbers of missiles to taper off toward a constant level. But if the world embarks on stage after stage of a great competition, the demand for actual weapons will be unlimited and the prospect will be of a large part of the available explosive stuff being used in an all-out nuclear war.

Some proponents of ABMs favor them partly in the hope that their deployment would lead to their further improvement, so that they might eventually be able to intercept most of a massive attack. Enigmatically, this would probably only make the situation worse for the noncombatants and no better for the combatants. Without ABMs, an all-out war between the nuclear giants might be over in hours, one or both sides having been beaten into collapse or submission by use of considerably less than total stockpiles. With very efficient ABMs, let us say ninety-nine per cent efficient or more as an extreme hypothesis, the war might take years as initial arsenals were first exhausted and then improvised means of delivery were substituted to use more and more of the explosive material that had not been made into weapons in advance. The nuclear giants might be destroyed as before, but in this case would drag the rest of the world down with them by means of fallout.

In summary, increase in number is a more menacing aspect of the future of nuclear weapons than development in kind, though both types of change will make the future more difficult to handle and intensify the threat to civilization. Familiarity has bred contempt, while the world and its statesmen have become inured to "living with the bomb" and fatigued with patient negotiations that have done little more than hide a lack of decision. Yet we have actually lived through only a very few years, too few to be significant as a precedent, of the small end of a nuclear arms race that will become even more overwhelming if rational steps are not soon taken to terminate it and to deflect the efforts of mankind into more constructive channels.

CONTESTS IN THE SKY

BY

ANDREW STRATTON
United Kingdom

Professor Stratton is head of the Mathematics Department at the College of Aeronautics, Cranfield, where he is concerned with the application of mathematics and computers in technology and management. He was formerly head of the Research and Assessment Group of the Weapons Department at the Royal Aircraft Establishment, Farnborough.

THE DEVELOPMENT OF WEAPONS, like any other technical field, depends on basic scientific knowledge, on fresh discovery, and on progress in technology. While some weapons (like the A-bomb) arise directly from a new basic discovery and require the development of novel and specialized technology, new weapons are not necessarily derived from new discoveries. The application of known technology in a novel way to the solution of new problems plays a major part in many developments.

The first basic step of commitment of science and technology to weapon application is the investigation, by theoretical and experimental studies, of the feasibility of some new weapon or component of a weapon. If feasibility is established and the military need exists, decisions are then required on the commitment of funds, which are a measure of the technological resources that will be used. Many developments involve very large expenditure. The means by which decisions are taken and resources committed are therefore as important as the availability or potential of a new technology. Priority in allocation of resources will be given to developments that promise high gains in military capability. Development and manufacture of specific weapon equipment may then follow, leading to deployment with the armed services.

The time scale for a major new development is long. For example, the possibility of nuclear fission was discovered and announced in January 1939. The first atomic bomb, which was really an experimental prototype, was dropped in August 1945. Full service deployment, and the development of the H-bomb, took another seven to ten years. From the discovery of a new principle to major weapon application can thus be a minimum of ten to fifteen years. The subsequent service life for a particular design of new equipment is likely to be ten years as a maximum; at the other end of the scale, the new weapon design may be out of date before it comes into service.

For these reasons, an important distinction has to be made between the forecast of potential new weapon developments, and the prediction of what weapons might be deployed at some time in the future. The former requires consideration of new discoveries and technologies that could have a major impact, of new developments that could take place by application of existing technology, and of technological problems that, if solved, would give a major increase in weapon capability. The prediction of what will happen in practice primarily requires a forecast of long-term allocations of major resources to meet predicted strategic and tactical situations.

In examining possible future developments in aircraft, missiles, and spacecraft, one must first analyze the fundamental factors that determine the effectiveness of the weapons. Logically an aircraft or missile, or a combination thereof, is a means of delivering the destructive energy of a warhead—the payload—to the target. The weight of payload and the range and speed of delivery required will determine the characteristics of the delivery vehicle. The destructive energy per unit weight of the warhead, the ways in which this energy is transmitted from the warhead to the target, the vulnerability to damage of the target, will all affect the warhead weight required to inflict some specified level of damage.

The over-all objective is to obtain the required degree of damage at a minimum cost. In past wars, the dominant contribution to the cost has been the loss of delivery vehicles as a result of the defensive or counteroffensive operations of the opposition; for example, the loss of bombers due to antiaircraft and fighter defences. With the complex weapons of today, however, the demands on economic and technological resources are such that the combined cost of procuring and operating the weapon—research, development, manufacture, manpower, spares, servicing, airfields, and so on—must be kept to a minimum. The final choice of a weapon is a complex balance between over-all cost and operational effectiveness, in relation to political, strategic, and tactical objectives.

It is convenient to consider some of the basic characteristics of the warhead in relation to the target before examining various possible types of delivery vehicle. I shall return to the question of decision-making toward the end of the chapter.

Warheads and Targets

For destructive purposes large amounts of energy have to be released in a very short time. The most efficient way of storing and releasing the energy within the smallest weight and volume is to use the primary chemical and nuclear energy sources. The table shows the amounts of energy that ideally can be obtained. The practical yield from nuclear weapons is less than the available energy quoted and is usually referred to as the equivalent weight in tons of TNT that would produce the same energy.

TABLE: *Amounts of available energy in primary sources*
The unit of energy used is the megajoule, which is equivalent to 1000 megawatts of power for one-thousandth of a second or to 280 watts of power for one hour; it is the amount of heat required to raise just over half a gallon of water from 10° C to boiling point. One kiloton (KT) of TNT equivalent corresponds to 4.2 million megajoules.

SOURCE	MEGAJOULES PER KILOGRAM	NOTES
Chemical high explosives		
TNT	4.2 ⎱	⎰ no external source
Nitroglycerine	6.2 ⎰	⎱ of oxygen required
Chemical fuels		
Petroleum	46 ⎱	⎰ require a source
Hydrogen	140 ⎰	⎱ of oxygen
Nuclear fission		
Uranium or plutonium	84 million	assuming complete fission
Nuclear fusion		
Deuterium	240 million	assuming complete fusion

Energy can also be stored in secondary power sources such as accumulators, and in intense electric and magnetic fields. Today, the storage efficiency of such sources, in terms of energy per unit of weight or volume, does not compare with that of primary sources, but for some applications this disadvantage is partly compensated by the readily used and controllable form of the energy—for example, electrical energy.

The writer of this type of article thirty years ago would have drawn attention to the large amount of energy that in principle could be obtained from the conversion of mass to energy in nuclear reactions; but he would have had to add that no means of releasing this energy on a practical scale had yet been devised. If some new basic discovery in the future produced a primary or secondary energy source with an available energy per kilogram weight intermediate between chemical explosive and nuclear energy, and without the odium and aftereffects of the nuclear warhead, then such a discovery would have a major military and political impact. Today, this must be considered as pure speculation and consideration will be limited to the exploitation of the known sources of destructive energy to obtain maximum weapon effectiveness.

The cost of delivering a warhead to the target increases with the weight. If no other considerations were to apply, then the choice between nuclear and chemical explosives would be made by balancing the greater cost of the nuclear warhead against the decreased cost of the means of delivery (missiles and/or aircraft), taking into account the destruction required and the likely losses due to enemy counteraction. The long-range ballistic missile, for example, is economic from the cost-effectiveness point of view only if it carries a nuclear warhead.

Immediately on its release, the energy of an explosive

appears as a combination of internal energy of a highly compressed gas, kinetic energy of motion of the products, and radiation energy. At the moderate temperatures of a chemical explosion the amount of radiation is small; at the very high temperatures attained in a nuclear explosion a high proportion appears instantaneously as radiation energy.

The ways in which energy is transmitted to the target, which may not be the same as the forms in which it is first released, are:

1. kinetic energy of a missile—for example, a bullet or metal fragment;
2. blast—vibration transmitted as a pressure wave through the atmosphere;
3. shock—vibration transmitted as an elastic wave through a liquid or solid;
4. nonnuclear radiation—ranging downwards in wave length from radio waves through infrared and visible light to ultraviolet and X-rays, the energy of the radiation increasing inversely with the decreasing wave length;
5. nuclear radiation—gamma-rays, beta-rays, and fast nuclear particles.

Commonly more than one method of transmission operates. A chemical explosive warhead, surrounded by metal, transmits energy by fragments, blast, and shock in proportions determined by design. A nuclear warhead exploded at low altitudes (below, say, thirty kilometers) transmits about fifty per cent of the energy in blast and shock, thirty-five per cent in thermal radiation, and fifteen per cent in nuclear radiation.

The use of rocket vehicles makes it possible to explode nuclear weapons at higher altitudes. The decrease in air density with increasing altitude reduces the effectiveness of the shock wave as a damaging agent. In the vacuum of space, transmission can only take place by kinetic energy of the explosive products and metal fragments, and by radiation. A major change takes place in the process of energy transfer from a nuclear warhead, at high altitudes. The radiation energy of the explosion appears initially mainly as soft X-rays; at relatively low altitudes this is immediately absorbed by the air, producing a large volume of high temperature gas—the fireball. The energy within this fireball is then propagated as blast and thermal radiation. At high altitudes, because of the lower air density, the X-rays travel much greater distances before being absorbed and thus heat a much larger volume of air (and, in spite of the lower density, a greater mass) to a lower temperature.

A high-yield nuclear explosion at an altitude of thirty to fifty kilometers will cause severe damage on the earth's surface by thermal radiation. At an altitude of one hundred kilometers there is a considerable decrease in thermal radiation damage. Any defensive action against large warheads must therefore be effective at sufficient altitude to avoid thermal radiation damage to the underlying territory.

At the altitude of engagement of a ballistic missile warhead by a defensive missile, the X-rays produced by a nuclear warhead carried by the latter will have a range of more than a kilometer. A one-kiloton warhead will give an X-ray energy density at 300 meters of about two megajoules per square meter; it will thus transmit to a thin surface layer of the target in a very short time energy equivalent to that produced by half a kilogram of chemical explosive, per square meter of the surface.

The vulnerability of a target to the impacting energy depends in a complex way upon the nature and duration of the energy and the characteristics of the target, which is not uniformly vulnerable. Targets may be "hard" (for example, a tank), requiring a high concentration of energy for damage, or "soft" (for example, a house or the human body). The degree of damage that has military significance also varies with the target. Virtual instantaneous and total destruction of a ballistic missile nuclear warhead by a defensive missile warhead is required, whereas in tactical warfare temporary immobilization of opposing forces may gain an immediate advantage for the attacker or defender.

The targets may be spread widely over an area (area target), in which case the destructive energy should be distributed as uniformly as possible over the area, or be individual targets (point targets), in which case as much as possible of the warhead energy should be concentrated on the vulnerable spots of the target. The accuracy of aiming or guidance, and the fusing to explode the warhead at the optimum point on the trajectory, are thus very important factors in determining weapon effectiveness.

For soft targets over an area—or where the position of point targets is uncertain—developments can be expected of means of controlled distribution and ignition of fuels over the area, which will give greater effectiveness than, say, napalm. As the table indicates, fuels that burn in atmospheric oxygen also give a much higher energy yield than the same weight of chemical explosive, which includes the necessary source of oxygen within its chemical composition.

The uncontrolled distribution of energy from a warhead is approximately uniform in all directions; if released at a distance from a point target, much of this energy is wasted. Significant developments have taken place in the concentration of warhead energy by controlling direction of fragmentation and by the focusing of blast energy by shaping the explosive charge. For example, the concentrated energy from a kilogram of explosive, focused in a jet, will penetrate thick armor plate. To be effective, however, any such concentration of energy has to be matched to the accuracy of aiming, guidance, and fuzing.

Considerable increase in effectiveness for a given warhead weight obtains if the energy concentration can be maintained as range increases. For example, a target area of one square meter at a distance of 1250 meters from a source giving uniform distribution will receive only one part in twenty million of the total energy available over the surface of the enclosing sphere. Conversely, if all the energy in one kilogram of explosive were concentrated on the one-square-meter target it would give the same energy density as twenty kilotons of explosive with uniform distribution of energy.

A direct way of achieving this high efficiency is to transport the explosive in a shell and detonate it at the target. But aiming and ballistic errors, and unpredicted motion of the target, make the chance of accuracy low, and thus reduce the over-all effectiveness. If it is feasible, there are obvious advantages in using the very high velocity and straight path of some form of radiation, as the method of propagating a directed, concentrated beam of damaging energy.

For attack on targets outside the atmosphere, and in order to obtain a high degree of concentration of damaging energy on the target, we can forecast increasing attention on the use of radiation as a damaging agent. X-rays produced by a nuclear warhead are the most likely way of destroying a ballistic missile warhead, and if some means of directing and concentrating the radiation can be evolved, this will increase effectiveness.

With the discovery of the laser, means of producing light beams with a very small angle of divergence and with very high energy delivered in a short time are now possible. The consequent danger to the eye from laser beams and the ability to make holes in metal are well known. A weapon that will blind the unprotected eye is technically feasible with current technology. As a minimum, the developing technology of the laser constitutes a threat against unprotected and naturally soft targets such as a satellite or an astronaut in a space suit.

To produce a weapon that damages harder targets requires a manifold increase in the energy density of the laser beam. The limitations of current technology lie in the large bulk of the secondary power sources that store the electrical energy and the low over-all efficiency of conversion from electrical energy to light radiation. Development of a feasible method of highly efficient conversion directly from a primary energy source (such as a rocket propellant) to coherent (laser) light would have a revolutionary effect in enabling the energy of the primary source to be transmitted instantaneously and (outside the atmosphere) without loss, to the surface of the target. The extent of damage will depend very much on the nature of the surface of the target. Protective measures, such as are used in any case on vehicles that have to re-enter the atmosphere at high velocity, can be employed to harden the target.

Missile Guidance

Automatic guidance requires some clearly identifiable characteristic by which the target can be distinguished from its background and neighboring objects (which may be alternative targets). Guidance may be "passive," operating on natural characteristics of the target such as infrared emission; alternatively, the target may be illuminated with energy (for example, radio waves) that is reflected from the target ("active" guidance). In the latter case, the reflecting characteristics of the target become an important factor in determining the accuracy of guidance.

Aircraft, missiles, and spacecraft have clearly identifiable characteristics (compared with their background) that can be used for guidance. Closely spaced real targets and decoy targets, however, create major problems in discrimination.

The ballistic missile, although unguided for the greater part of its flight, relies on guidance equipment to give it an accurately determined velocity at a particular time and place after launching, so that subsequently it follows a predicted path with the requisite accuracy. The future position of the target relative to the launch point must be predictable at the instant of launch. Inertial guidance is the method commonly used; it involves a combination of highly accurate gyroscopes and instruments for measuring acceleration. From the military point of view, inertial guidance has the very major advantage over radio or other methods in that there is no known way of interference by external means; in other words, it cannot be jammed by the enemy. While scientifically the principles are simple—a direct application of Newton's laws of motion—the development of a technology that will give adequate accuracy has required the application of large research and development resources. When a mobile launcher is used (for example, the Polaris submarine) the position at launch has to be known accurately. Inertial navigation is also the key to navigation of the submarine.

The ease with which the laser can produce a beam of radiation of very small angle of divergence from a small area of illumination is of considerable importance in obtaining accurate guidance to a target and can be expected to produce a guidance system that will give a high chance of obtaining a direct hit on an individual target. Within the lower levels of the atmosphere the laser's range will be restricted because of atmospheric loss, but at high altitudes and outside the atmosphere no such restriction obtains.

The laser also offers prospects of producing a gyroscope for inertial navigation of very high accuracy, with no moving parts to wear out, and almost unaffected by vibration and acceleration. The principle—the change in velocity of light in a rotating frame of reference—was first demonstrated in 1913, but the sensitivity of measurement was too small to be practical. Recent work has shown that an increase in sensitivity of more than a million times can be obtained by the use of the highly coherent light of the laser, and although there are still difficult technological problems to be solved the laser gyroscope is likely to bring a major increase in accuracy—and, hopefully, a reduction in cost—to inertial navigation and guidance.

In communications, again particularly outside the atmosphere, lasers give the ability to transmit large amounts of data, securely, between vehicles.

Aircraft versus Guided Missiles

The aircraft may be considered as a mobile weapon launcher that can be recovered and re-used. It carries, however, one or more human beings who can analyze a tactical situation and recognize and acquire targets. It will be a considerable time before, in the attack on ground tactical targets, these functions can be performed as effectively by artificial aids. To perform these functions the human being and an expensive aircraft have to be exposed to risk; the outcome of the duel between attack aircraft on the one hand and defensive missiles and interceptor aircraft on the other is a major factor in predicting the future of strike aircraft.

Antiaircraft guided missiles are a very effective defence against aircraft within their field of fire. Coverage at high altitude is limited by missile performance (a matter of size and cost rather than technical limitation) and at low altitudes by complication of surveillance and guidance problems. The missile designer has to arrive at a compromise based on estimates of the likely threat, and the aircraft can counter by flying above or below the coverage until the missile system is extended to meet a new threat. The U-2 reconnaissance aircraft was a particular example of this interaction of aircraft and missile defence.

Interceptor aircraft armed with air-to-air guided missiles are more flexible in usage and deployment than are the long-range ground-based antiaircraft missiles. Against low-altitude targets there are major problems for the interceptor in obtaining adequate warning of attack (special radar surveillance aircraft are required) and in missile guidance.

Two roles can be seen for the interceptor in the future—to counter an enemy threat outside the coverage of ground-based guided missiles, and for interrogation. The latter is a most important role in avoiding serious international incidents. The interceptor, given the necessary speed superiority, can close for visual identification and take warning action against the intercepted aircraft; the launch of an antiaircraft missile, however, implies positive identification (which may be erroneous) and a decision to attack and destroy the target.

The over-all limitations in the cover provided by interceptor aircraft or guided missiles lie in the economic choice of an upper-altitude limit, and the technical problems of surveillance and guidance at very low altitude. Air strike is thus being forced to either very low or very high altitudes, to escape interception. Against tactical ground targets, the need for visual acquisition of the target and accurate weapon delivery has so far resulted in a concentration on very low-level operation.

The development of special low-level antiaircraft missile

defences, against which high speed at low altitude is no longer a protection, then forces the strike aircraft to avoid these defences by launching a "stand-off" weapon from outside the defence coverage. Against ships and radars, automatic guidance methods can be used, but for tactical ground targets it is at present necessary to continue to use the human operator, linked remotely by television. One can conceive of highly sophisticated means—perhaps based on small temperature differences—that will enable certain types of tactical targets to be distinguished against a confusing background, and of complex and highly compact computers that will analyze the responses and automatically direct a weapon to the target. Such weapons would inevitably be costly and economically viable only if used against major targets with a warhead of much greater effectiveness than present chemical explosives.

For air attack on military targets in close support of ground troops, a likely development will be to bring guided missiles, whether launched from the air or the ground, under the direct control of the soldier on the ground who has located the target. The development of highly reliable and compact equipment will allow full exploitation of the capability of VTOL (vertical takeoff and landing) aircraft to operate in the field. The development of cheap, lightweight equipment will also give small helicopters and hovercraft the ability to move and fight by night; the use of natural cover will give some protection against defences.

The extent to which low-level aircraft operation over enemy territory is made impracticable by the low-level defences is primarily a matter of the cost of the large number of defensive units necessary to cover an area. The effective range of a low-level defence is limited by terrain screening the target. But the advent of a cheap lightweight antiaircraft guided missile that could be used in large numbers by the soldier in the field would weigh heavily against the strike aircraft operating over enemy territory.

Against a less well-equipped army, the limitation in air strike is likely to continue to be the ability to recognize and identify the target; increased development and use of area attack weapons is the only foreseeable trend.

As the area of coverage of the defences increases, the necessary "stand-off" range of the strike aircraft's missile increases, with consequent increase in its size and cost and decrease in the aircraft's effective load. An alternative counter is to overfly the defences—that is to say, "stand-off" in altitude. While this is technically feasible, it requires a costly aircraft that would be able to carry only a very limited load of missiles; and to be viable it would require nuclear warheads.

The fundamental question is thus whether, against an enemy equipped with air defences that are both foreseeable and technically practicable today, the strike aircraft will be an effective weapon in land tactical situations, without resort to stand-off attack with nuclear warheads.

Moreover, the same question arises when one considers manned interceptor aircraft. With the need to have speed superiority over the supersonic civil aircraft now being designed, the next generation of interceptor, equipped with long-range airborne radar, infrared detection and guidance equipment, compact computers for tactical evaluation, and highly accurate air-to-air guided missiles, will be very costly to develop, manufacture, and operate. A decision to produce such an aircraft would thus be unlikely unless some other role—for example, as strike or reconnaissance—could be foreseen. Yet, as we have just seen, in a strike role the effectiveness in relation to cost would be low unless nuclear warheads were used.

The possibility of evolutionary extension of civil aircraft speeds to Mach 5 and heights to thirty kilometers and beyond emphasizes the question of whether manned interception is economically viable and, if so, whether it would take place from earth or from a vehicle in orbit.

Ballistic and Antiballistic Missiles

The primary element of ballistic missile technology is rocket propulsion to give the payload the requisite velocity—about 6.5 kilometers per second for an intercontinental ballistic missile (ICBM) of 8000-kilometer range. Other basic elements are a guidance and control system of accuracy, matched to the range, warhead yield, and target characteristics; and a re-entry body that can bring the warhead and fuse safely back into the atmosphere at high speed. With current technology it is practicable to deliver a nuclear warhead of any conceivable yield from any point on earth to any other point, and current and foreseeable guidance developments (such as the laser gyroscope mentioned earlier) will enable the accuracy to be matched to the characteristics of the target and the yield of the warhead.

An antiballistic missile defence, like any other defensive missile system, incorporates:

1. surveillance, target tracking, and acquisition;
2. missile preparation, launch, and guidance;
3. terminal guidance, target interception, fusing, and warhead.

The special technological problems arise from the distances and high speeds involved and from the high degree of effectiveness required, bearing in mind that the ballistic missile warhead is an H-bomb.

Consider, for example, defence against an ICBM of 8000-

kilometer range. The time of flight will be less than thirty minutes, and the peak of the trajectory (apogee) about 1100 kilometers above the earth's surface. The surveillance problem is to detect and identify the launch of such a missile, as soon as possible. Because of the curvature of the earth, land-based radars must be placed as far forward as possible; very high powers and large aerials are needed. With the large volume of space that has to be kept under surveillance and because of the presence of many radar targets, very complex and sophisticated computing techniques are necessary to keep the rate of false alarms low; such analysis requires repeated reception of reflected energy from the target and takes time.

Once the presence of a new target is recognized, it has to be observed for sufficient time to determine its trajectory and speed. The accuracy of range, bearing, and elevation given by surveillance radars is limited, because the time to search a given volume increases as the accuracy requirement is increased. Therefore, when a threat has been identified, an individual high-precision tracking radar beam (which will probably be another radar) must be allocated to it; as the surveillance radar can give only the approximate position of the target, further time elapses while the tracking beam is searching for and acquiring the target. The tracking beam obtains accurate position and velocity information from which predictions can be made of the complete trajectory for interception purposes.

The target-handling capacity of the defensive system will be determined by the number of tracking radar beams. Although variations in the actual radar techniques used can be expected, there is no escaping the fundamental limitation that the accuracy of information on the target and the rate at which it can be obtained depends on the amount of reflected energy. While development of satellite surveillance techniques may enable missile launches to be detected—for example, by infrared emission—the acquisition of the target by a tracking radar will still be necessary.

The defensive missile (ABM) itself poses many problems. The target may have covered half the trajectory by the time the ABM can be launched; a short reaction time for the launching sequence and very high acceleration to a speed comparable with that of the target are essential. The only form of control that can be used outside the atmosphere is variation in the direction of thrust of a rocket motor, and consequently the ABM will have to be powered throughout flight; even so, maneuver will be limited by the available thrust. The flight path requiring the least maneuver is a collision course to a predicted interception point on the ballistic missile's trajectory.

The missile launching site is likely to be thousands of kilometers from the surveillance radar, and tracking of the target at some stage will have to be handed over to a radar at the launching site. To guide the ABM on a collision course, accurate information is required on the ABM's velocity and position; this necessitates yet another radar to track the ABM, and possibly inertial guidance equipment in the ABM itself.

A most difficult guidance problem arises in identifying the actual target—namely the re-entry vehicle carrying the warhead. Separation from the launching rocket will probably have occurred, and, apart from the false targets created by parts of the rocket, decoys will have been released to confuse the ABM guidance system. Identification of the true warhead may be possible by delaying interception until either the trajectory can be identified or the ionization and infrared emission produced on re-entry can be detected. But such delay brings its warhead dangerously close to its target.

Even if the true target has been acquired, the very high closing speed means that little maneuver is possible. For example, the distance between ABM and target will decrease by about thirty kilometers in the time taken for a sideways thrust of ten times the ABM's weight to deflect the ABM only 300 meters to one side of its original track.

Accuracy of guidance is difficult to obtain with target and missile being tracked by separate radars. As soon as the interception trajectory permits it, a terminal guidance phase begins, in which the relative trajectory between ABM and target is determined as directly as possible. Direct homing by the ABM itself onto the target would give the greatest accuracy, but the long range from which homing would have to start and the problem of acquiring the target by a suitable homing device present formidable technical problems.

Relatively large miss-distances must therefore be expected, and a nuclear warhead will be essential for the ABM. Its yield will be chosen in relation to the expected miss-distance and the hardness of the target.

The severity of all these technical problems is only too evident. Although developments in electronic scanning of radar beams, signal extraction, data handling and processing, and in missile propulsion, guidance, and fuzing, will give enhanced performance, only a revolutionary development —such as the means of controlling the direction of emission of high intensity X-rays from a nuclear warhead— could give a major increase in the effectiveness of ABM defence.

The problem of ABM defence is, however, as much strategic and economic as technical. Saturation of the defences by sheer numbers of missiles and by decoys is the obvious counter. The system must be designed on the basis of pessimistic predictions of its effectiveness and the number of targets; there will be no second chance to eliminate deficiencies in design or deployment. The cost of defending against ballistic missile attack, even over a limited front, is extremely high. With the possibility of much larger missile ranges, a limited defence can be negated by attack from other directions, launched from land, sea, or space.

Satellites and Antisatellite Systems

The existing ability, illustrated in manned spaceflight, to put a large payload into orbit and then to obtain controlled re-entry means that in principle a similar technique could be adopted with an H-bomb. Attack could then be launched around the earth, avoiding radar and ABM screens. Alternatively, warheads could be "parked" in orbit for re-entry on command. While large warheads could be exploded in orbit without warning, with diffuse heat damage to the underlying territory, their effectiveness would be considerably increased by re-entry toward the target.

The satellite missile would have to contain a propulsion system to reduce the velocity for re-entry, and a means of guidance. As errors in inertial guidance accumulate with time, the guidance requirements are more stringent than for the direct ballistic trajectory; if necessary, automatic star tracking could be used to supplement improvements in inertial guidance techniques. Provision of a ground communication link that could not be rendered inoperative by jamming, with a very high degree of security against malfunction, and a high probability of operation when required, would be the major problem in a command system.

Sufficient warheads would have to be placed in orbit to cover the desired targets and phased so that a nearly simultaneous strike could be delivered. Operating in an orbit lower than the apogee of an ICBM, the re-entry satellite takes less time to leave orbit and reach the ground than the ICBM descending from its apogee. The greater part of the already meager ABM defence reaction time would thus disappear. Even if all suspect satellites were tracked continuously by the defence, the chance of interception by ABM in time to avoid damage from a high-yield warhead would be small. Any malfunction during placing in orbit that resulted in a warhead detonation, or any accidental re-entry, could trigger a nuclear counterstrike.

The situation created by the deployment of nuclear warheads in continuous orbit would be most hazardous and invite immediate counter against the satellites and against further launching, including counterstrike. To detect the initial deployment of such a weapon amid frequent civilian satellite launchings would be difficult, and opens up the question of satellite inspection.

Satellites have other military uses. Unmanned satellites have already demonstrated their capability in the communications and meteorological surveillance fields, and the present ability of the satellite to "overfly" all defences gives it considerable potential for military reconnaissance. High-resolution airborne reconnaissance aids, based on millimeter radio waves or infrared or optical methods are applicable to satellite operation and will continue to improve in resolution and accuracy. A very large amount of information is collected in each transit, and rapid communication to the ground and rapid analysis are essential. Developments of automatic pattern recognition can be expected which, by employing large computer capacity, will identify changes in reconnaissance data. Continuous surveillance of all launching areas for missile and space vehicles will have to be maintained if any significant contribution is to be made by satellites to ABM defence.

Satellites themselves may be attacked by an ABM type of missile; in their present mechanical form and with solar cells for power they are very soft and vulnerable targets, and their orbits are accurately known. Ejection of decoys, self-contained nuclear power sources, and general "hardening" and protection of the satellites are the most direct counters.

Any assessment of the military potential of manned satellites and spacecraft raises many questions. Compared with the case of unmanned satellites, a high price has to be paid for providing conditions in which men can survive and work in space. A man can act only at a slow rate compared with that of a servomechanism; he takes in information relatively slowly, and compared with an electronic computer he is slow and inefficient in processing large amounts of data. The value of man in space lies in his ability to discriminate between objects and to identify and classify them, to analyze an unpredicted situation, to exercise judgment, and to react to the situation. Unmanned satellites, however complex, can operate only in ways that have been programed before launch, and will be rendered inoperative by failures that have not been foreseen and guarded against by redundancy. A manned satellite can, within limits, adjust and reprogram the operation of the equipment to meet unpredicted events, and identify and repair failures.

An extension of the satellite warhead concept would be to have re-entry of the warheads from orbit controlled by a manned satellite. By using laser links for control, a high degree of security could be maintained and a nuclear strike could be launched irrespective of any conditions prevailing on the surface of the earth.

The inspection of satellites for possible nuclear warheads, so as to allow prompt neutralizing action, would require manned spacecraft with sufficient propulsive power to give adequate maneuvering ability.

The need to provide a local environment to sustain life renders manned spacecraft very vulnerable, and considerations of size and weight would make hardening difficult. A nuclear warhead on an ABM type of missile would have such a large lethal range that escape by maneuver would be impracticable; reliance would have to be placed on decoys and other countermeasures directed at the ABM guidance system. A manned spacecraft would, however, be vulnerable to much simpler methods of attack; a single direct hit on the crew compartment from a small high-explosive shell would be lethal.

As seen from a neighboring spacecraft with low relative velocity, the craft-to-craft attack problem is, if anything, simpler than on earth. In the absence of drag there is no restriction on propulsion range (although a small thrust motor would be required for control); the effective gravity is zero, although there will be a small (Coriolis) acceleration owing to the orbital rotation that will curve the flight path of the missile relative to the spacecraft. Some form of guidance might be needed, but simple visual command guidance techniques—for example, as used in antitank weapons—would be adequate. For more difficult situations requiring a high missile velocity, a high-energy laser could become the spacecraft-to-spacecraft weapon.

Automation in Advanced Weapons Systems

It is commonly supposed that automation removes the possibility of human error; on the contrary it transfers the possibility of human error from the point of action to the design and development process. Errors and malfunctions in equipment are bound to occur, and the human designers exercise their individual ingenuity in predicting and guarding against the consequences of failure. We are just reaching the stage at which the possibility of an accident in the few critical minutes of landing an aircraft completely under automatic guidance is less than when a human pilot is in control.

It is questionable when, if ever, the stage will be reached where a nuclear weapons system can be left to complete automatic control, with less certainty of error than ultimate human decision. Yet possible future developments, notably in missile exchanges, leave little, if any, time for introducing human decision and control into the system.

The Choice of Weapons

Current technology has already advanced to the stage at which many more possibilities exist for weapon development than are economically feasible even for the most advanced nations. Even without revolutionary developments, the general evolution of technology will widen the choice of possible lines of action and increase the sophistication and cost. The brief survey in this chapter has discussed some of these possibilities. Whether or not such weapons will be produced depends on assessments of the gain in military effectiveness in predicted strategic and tactical situations, on the availability of resources for the development, manufacture, and deployment of the weapon, and on political decisions.

That defence is an economic as well as a technological and strategic problem became evident in the nuclear deterrent program, and techniques were evolved for evaluating the effectiveness of a new weapons system in relation to total cost—cost-effectiveness analysis. Such analysis requires study of the technical feasibility and performance of new weapons systems, an assessment of their military effectiveness, and an estimate of the total cost. Although based wherever possible on experimental evidence, by nature much of the prediction of performance depends on analytical estimates and "paper" design.

The assessment of military effectiveness has to be based on mathematical models and supporting experimental evidence of encounter between warhead and target, and on a mathematical model of interaction between offence and defence. Both require explicit assumptions and predictions about the future equipment and actions of a potential enemy. The estimate of true cost is subject to all the uncertainties of advanced technological development and to future changes in political and strategic policy. For major weapons systems the analytical models are extremely complex, and the interaction between offence and defence may be extended, in "war games," to include the effect of decision-making by either side.

The object of the analysis is to present the interaction between choice of weapon, military effectiveness, cost, and an inevitable range of assumptions from predicted technological success to enemy strategy and tactics, in a sufficiently concise form for high-level decisions on weapon policy and development to be taken with full knowledge of the military and economic consequences. As the scale of the problem grows, increasing reliance has to be placed on an analysis giving a supposedly unbiased representation of the interactions investigated.

The difficulties of producing an unbiased representation, and the consequent dangers in the analysis, lie in the many assumptions that have to be made by the many persons involved. Although made with all intellectual honesty, these assumptions will reflect at all levels the fallibility of human judgment and may not even appear as specific assumptions in the final analysis. With the increasing complexity of the weapons system there is thus the growing possibility that the effective judgment in the decision is not that of a few at the highest level, but that of many at a much lower level.

Difficult and imperfect as the cost-effectiveness analysis process may be, weapon developments have become so complex and costly that unaided human judgment of the few may be less likely to lead to a viable policy.

From the military viewpoint the interaction between offence and defence is the dominant factor in weapon selection. Defensive measures may be purely passive (as in the ABM) or counteroffensive (as with the nuclear deterrent). The design characteristics of a passive defensive weapons system have to be determined by an estimate, some seven to ten years in advance, of the technical character of the offensive weapons against which it will be deployed.

Economic considerations will limit the scale of defensive deployments and determine a level of saturation above which any attack can penetrate unmolested. The offence can use unforeseen tactics and countermeasures against the defence and achieve a surprise initial penetration. This interaction may be seen as an economic conflict between the two contestants. For example, wide deployment of a cheap and effective low-level antiaircraft defence can force the offence to considerably greater expenditure on stand-off weapons—or to accept higher losses. Again, an expensive and complex ABM system may not be economically viable if it can be countered at a much lower level of expenditure by the deployment of more ballistic missiles, or if it inhibits other developments giving a greater return in over-all military capability.

Many of the possibilities discussed in this chapter, if pursued, will lead to an even heavier commitment of scientific and technological resources to military purposes than ob-

tains at present, with a higher proportion of each participant nation's economy growing dependent on the continuity of the development. There are obvious close links,
in technological development, with the civilian space programs. Space research and the manned exploration of space
provide a convenient reservoir for absorbing effort from,
and providing effort for, a fluctuating defence program
and advance technology in areas of direct application to
new weapons. The challenging nature of the problems ab-
sorbs the interests of scientists and engineers to the neg-
lect of other developments. Although the advancement
of technology and "spin-off" to nonspace and nonmilitary
applications is cited as a worthy objective in itself, a fraction of the resources directly applied to these other fields
would give no less an advance.

Any limitation or control of future weapon developments
in the field of missiles, satellites, and manned spacecraft
will thus have to extend also to the control and definition
of objectives of national space programs.

THE STRATEGIC
CALCULATORS

B Y

HARVEY WHEELER
United States

Dr. Wheeler is a fellow-in-residence at the Center for the Study
of Democratic Institutions in Santa Barbara, California. He previously taught political science at Harvard, Johns Hopkins, and
Washington and Lee University. He is co-author, with the late
Eugene Burdick, of the novel Fail-Safe.

COMPUTERS have a direct effect in three immediate aspects
of warfare: 1. the control of weapon performance; 2. policy
formation and decision-making; and 3. the administration
of military programs and operations. All are relatively restricted areas, but from them, singly and in combination,
spring far-reaching ramifications whose ultimate consequences are impossible to foretell. In the first parts of this
chapter, I shall describe the present significance of computers in the American military establishment—a signifi-

cance not widely appreciated—and in the latter parts I shall discuss the implications for warfare in the future.

We are dealing with an innovation whose effects promise to be comparable with those that followed the introduction of double-entry bookkeeping, in Genoa in the thirteenth century. Then the immediate effect was to permit the expansion of the commercial transactions of a business beyond the capacities for monetary management of an individual proprietor. A much more ancient invention, the corporation, had released the *scope* of organization from dependency upon the coordinating and operating abilities of the individual. Yet it was only in the sixteenth century that both features were combined to produce the business enterprise as a separate economic entity, distinct from its owner or owners. No thirteenth-century Genoese merchant could have perceived that one of the technological sources of an industrial revolution was embedded in his newfangled bookkeeping system. Similarly, no contemporary observer can possibly imagine what developments will ultimately spring from the crude calculating device that a few wartime scientists developed in the 1940s.

The first assignment of computers was merely to solve the complex simultaneous equations required to bring about an intersection between the flight path of an approaching enemy aircraft and the trajectory of antiaircraft artillery. Immediately, however, a revolution in warfare was set in motion.

Performance of Weapons

The problem of increasing the accuracy of antiaircraft fire involved programing a complete system with three types of components linked together: sensory input equipment (radar), a problem-solving device (computer), and a motor response (weapon). Modern "systems engineering" was born. The obvious if crude parallel to the functioning of the human being was immediately noticed by men like Norbert Wiener, who coined the word "cybernetics" to cover the complete range of biological, mechanical, electronic, and social systems that appear to operate according to the same general systemic principles.

Soon the problem of directing antiaircraft fire was replaced by the problem of guiding missiles, both for offence and for defence. As the more sophisticated intercontinental ballistic missiles (ICBMS) and space vehicles were developed, more advanced sensory and electronic machines were also produced, leading to "real-time, on-line" weapons systems. Missiles and space vehicles could not be guided with the accuracy now possible if we did not have the contemporary high-speed computer. Similarly, defence against ballistic missiles could not be contemplated if advanced computers were not available to control and guide the necessary systems. There may be, at best, only fifteen to thirty minutes' warning that an enemy missile is approaching; action and reaction must both occur in the same time phase. In the past, for any type of warfare, a counterstrike did not have to occur at exactly the same instant as the inaugural strike. A threatening battleship detected one day would still be there to receive a counterattack the next day, or even later. There could even be enough time to make official inquiries or protests through the traditional channels of diplomacy. But the blinding speed and the cataclysmic destructiveness of contemporary offensive weapons requires the "real-time" deployment of defensive weapons, if they are to have any effect at all.

The practical meaning of a "real-time" system of control depends upon the time span within which perception, decision-making, and action, the three elements of the complete system, may be operated effectively. Compare other types of time systems. A philosopher like Immanuel Kant, deliberating speculative problems posed by David Hume, operated in a time span that could occupy generations. For a politician like Charles de Gaulle, pondering how to make the diminished power of France effective under the novel conditions of the Cold War, his effective time span may comprise a whole year, or even a decade. For the pit crew of the lead race car at the Indianapolis speedway, the effective time span may be less than twenty seconds. For a missile defence system, real time is closer to three millionths of a second. This kind of time scarcely existed at all, in any meaningful sense, for previous generations. Now, the need for instantaneous analysis and control is showing up in all sorts of circumstances, from complex manufacturing processes to traffic control at busy airports. An ever-increasing variety of tasks require something like the split-second analytical capacity and the high accuracy pioneered in missile control systems.

Two types of control systems are possible: DDC (Direct Digital Control) and SC (Supervisory Control). Under DDC the system is completely computerized. Under SC there is an intermediary stage between computer and motor response in which human intervention may occur. Most complex systems provide elements of both. A missile defence system, for example, possesses DDC components, but the over-all system is SC, because the final decision is preserved in the hands of the Commander in Chief. Where a system involves discrete operations that may be irreversible, like the launching of a nuclear-armed missile, some of the speed of response offered by DDC should be sacrificed to insure certainty about the information presented; in such cases, human intervention through an SC system is obviously required.

The difficulty is that the human being may prove to be the system's undoing. If the information is accurate, yet unbelievable, a man tends to distrust the computer's analysis and substitute his own. Computerized calculations of collision courses for two moving objects often appear erroneous to the human observer. A man, seated at his console, monitoring on radar the path of an incoming missile and the deployment of its interceptor, often feels like the operator of a pinball machine. The computer-directed interceptor doesn't seem to be taking the right path. The monitor has an almost irresistible impulse to "tilt" the machine a little and "correct" the interceptor's direction of flight. At least one airline disaster is thought to have resulted from a pilot's belief that the evasion course given him by the control tower was in fact sending him on a collision course with a nearby airliner. It was the substitution of his own judgment that brought disaster. Here is the dilemma of the human intermediary in a computerized system. He knows that disbelieving the computer, even when it tells him something that defies belief, may bring disaster. Yet, when it is an intercontinental ballistic missile that is about to be launched, the discord between computer analysis and human perception is such that human action, or inaction, takes place in what is, in effect, a state of human ignorance.

In all such computerized systems where instantaneous motor response is necessary, but where the effects of the response may be irrevocable, the most obvious remedy is to build into the system redundant, back-up components. These are duplicate or supplementary information systems to insure that there has been no failure, or error, in the primary system. In precomputer control systems redundancy was synonymous with inefficiency. It was the bane of all efficiency engineers. One of the strange accompaniments of the advent of computerized systems is to erect into a principle of efficiency the very feature that had been its contrary in the past.

An example of an auxiliary information system is found in the "hot line," the ever-ready teletype link between the White House and the Kremlin, which has become the necessary diplomatic adjunct of computerized nuclear weapons systems. Moreover, decentralized ambassadorial diplomacy has tended to give way to more direct, chief-to-chief, diplomacy at summit conferences. No significant international issue has been resolved at any Cold War summit conference, but it provides an occasion for those who may have to activate culture-killing computerized weapons systems to make personal assessments of the people at the other end of the hot line.

Command and Control

The second way in which computers have influenced warfare occurs in "command and control," which refers to the broad area of military strategy and decision-making. It applies to the immediate problems facing regimental commanders in the field as well as to some of the more general problems facing the highest authorities.

A typical example comes from a rather straightforward allocational problem. An aircraft is lost at sea. A limited number of search planes must be given search patterns that will maximize the chances of finding the downed pilot within his estimated survival time. From an analytical point of view this particular problem is relatively easy for a mathematician, if not a field commander, to solve by paper-and-pencil methods. Yet when such a problem is magnified a thousandfold, only sophisticated computers can provide the solution. An example might involve the manned bombers of the U. S. Strategic Air Command. How many, flying what patterns, with what frequency, are required to maintain an ever-ready striking force?

Further removed from the theater of operations are longer-range questions concerning the allocation of funds for proposed new weapons systems. One must choose between a medium-range, high speed fighter-bomber or one with a longer range but less versatility. One must decide whether to rely upon nuclear warheads in rapidly activated, very fast ICBMs or to maintain manned bomber delivery systems that are slower but provide better safeguards against accident or miscalculation, as well as a more flexible range of response.

Here we are dealing with an approach to military problems similar to what is sometimes called the "cost-benefit" approach. Use of this approach does not necessarily depend upon computers. In fact, it was developed from relatively simple types of economic analysis, such as those used when a business firm chooses between alternative investments by comparing their estimated rates of return. However, as a practical matter it is almost impossible to apply cost-benefit analysis to complex problems or operations— as abound in the war offices of the large nations—without computers. In a well-publicized directive President Johnson required all U. S. Government departments to institute the cost-benefit approach to all possible decision-making problems. Computers eliminate the gigantic clerical operation that this approach would otherwise entail, and make it possible to engage in certain kinds of analysis that would otherwise be inconceivable.

This consideration brings us to "systems analysis." Although systems analysis usually refers to systems involving men, computers, and machines, it is not necessary to employ computers to do systems analysis. Operational research was something like systems analysis and it predated the computer. Important refinements were made at RAND, the "think tank" sponsored by the U. S. Air Force at Santa Monica, California. These refinements were produced primarily by social scientists rather than by computer experts and mathematicians, but the influence of the computer should not be discounted. One thing the computer does is to force anyone who would make use of it to think clearly, to formulate goals precisely and to express them unambiguously—which may indeed prove to be the computer's greatest boon to man.

One implication of the systems analysis must be pointed out immediately. It has been the device whereby a *coup d'état* was executed under Secretary Robert McNamara in the U. S. Defense Department. The Defense Department had been organized to eliminate the parochialism of the separate armed services and to reassert the principle of civilian control of the military. Systems analysis—and the related method of "planning-programing-budgeting"—was the device through which the accomplishment of this aim was finally set in motion. McNamara, buttressed with his elaborate cost-benefit analyses, was able to fight off the self-interested demands of the separate services and centralize ultimate decision-making in his own office.

However, this accomplishment was not without its costs, and the military services charged the systems approach with conservatism. The case was most persuasively presented by the American military curmudgeon, Admiral Hyman Rickover, concerning McNamara's opposition to nuclear-powered surface warships. Rickover's argument was that, if one follows the cost-effectiveness approach, it is necessary to take supplies, facilities, and fleet conditions as they exist and then project the comparative costs of refining present types of equipment as compared with the ramified costs of instituting entirely new naval systems. This procedure necessarily loads the calculation concerning any drastic innovation with a host of costly subsidiary items entirely absent from calculations concerning the maintenance of things pretty much as they are. The result is a built-in conservatism whenever fundamental innovations are at issue. This, as we shall see later, is a kind of conservatism that runs to the essence of decision-making.

By modern communications, command and control operate directly from the highest headquarters to the battlefield. Authority is concentrated at the highest command and control level; the analytical component of top-level decision-making is expanding as the element of intuition, pragmatism, and experience diminishes in significance.

In the past battles were fought for wars that had already ended and truces were celebrated for those that had not. Leaders waged war much as children played blindfolded at pin the tail on the donkey. Although the city of Washington lies only a hundred miles or so from Gettysburg, President Lincoln was no more in control of the crucial battle there than if it had been fought in Istanbul. All he could do was to send a prebattle morning wire to General Meade requesting to be informed of the outcome that night.

Today, information is not only instantaneous, it is also copious beyond belief. A so-called intelligence community consisting of eight intelligence organizations has headquarters in and around Washington. Each of the armed services has its own intelligence branch. Several separate agencies, such as the Treasury and the State Department, have their own intelligence bureaus. Then there are the F.B.I. and, most particularly, the C.I.A.; the Central Intelligence Agency is reputedly far and away America's largest governmental agency. The world-wide operations of these agencies funnel masses of information into Washington. Data-processing on the present scale was impossible before the computer; even if as much information could have been collected, it could not have been processed by paper-and-pencil methods within the time span relevant to the decisions on which it bore. Nowadays the President can be given, each morning, a composite, up-to-the-minute intelligence report on every critical event in the world.

The President, as Commander in Chief, thus has more complete information than can be acquired by anyone else in his entire system of command and control, including commanders on the spot. The first result has been the progressive sublimation of tactical considerations into the strategic sphere. Given sophisticated techniques for gathering and analyzing information, the discontinuity and discreteness once possessed by remote events gives way to an over-all interconnectedness. For example, a small-scale air strike by American bombers on Hanoi might be related to a whole series of larger problems of which no field commander could possibly be aware. A bombing mission could affect the internal strife between belligerent and conciliatory factions inside the Hanoi regime. It could also bear upon Soviet decisions on the quantity and quality of assistance to the Viet Cong. It could also affect Peking's calculations as to whether the United States had aggressive designs on China herself, and hence Chinese decisions about whether to enter the Vietnamese war directly. It could influence Britain's decision whether or not to support the American effort. It might reactivate the protest movement inside the United States, and so on, through a vast network of ramifying issues.

As was widely reported in the press, President Johnson personally pinpointed each day's bombing targets in Vietnam and received the same night direct reports of the results. This centralization of decision-making takes place not just because it is now technologically possible, but because, being technologically possible, it has become politically necessary.

In previous times the same events could not have had the same strategic or diplomatic implications because they could be discounted as the autonomous, unwitting acts of local commanders. Now, however, each side must assume that the other is aware of the secondary implications of all military actions. This means that each is forced to assume responsibility for the secondary as well as the direct effects of every military operation. Authority, in such a case, follows responsibility. The Commander in Chief must take a direct decision-making role in remote and seemingly minor tactical operations.

The process that converted tactical problems into strategic ones did not stop there. It went on to convert them both into diplomatic problems. This is one of the most revolutionary of the host of novel developments that has grown out of computerized warfare. With the outbreak of war between Israelis and Arabs in 1967, President Johnson sat at his White House controls with one hand guiding his Sixth Fleet and the other on the hot line to Moscow. The integral nature of these functions was verified when an Israeli attack on the American intelligence ship Liberty forced activity by American carrier-based aircraft. Moscow was also monitoring every radar blip, just as was Washington: would the Russians interpret the changing and converging flight patterns as an act of aggression? This was where the hot line came in, as Washington immediately explained its actions and reassured Moscow.

In previous times the general, the diplomat, and the statesman employed different tools and plied different trades. Computerized warfare has congealed war, diplomacy, and politics into one overarching art. Clausewitz's famous dictum has been superseded. War is no longer the continuation of politics by other means. Both now take place in a simultaneous synthesis.

Administration and Management

The third set of developments in computerized warfare concerns administration and management. This does not necessarily mean the over-all management of war. Rather, it refers to the management of such things as weapons development programs, logistical support operations, and problems associated with organizing and training troops. Similar problems occur in the management of large civilian operations. If it is necessary to know how many Portuguese-speaking licensed ham-radio operators there are in the Air Force, or to know almost anything else, the answers can be obtained quickly. This provides a capacity for management in detail that would not be possible without computers. It has, in effect, created a new field of activity, which might be called "support management."

For example, it was always known that, in order to support a regiment in the field, certain back-up services and supplies were also required. However, the information was terribly imprecise, and troops in the field were always plagued by supply imbalances and bottlenecks. This was partly because needs simply could not be accurately analyzed and calculated. Nowadays, in sending a fresh division into battle it is possible to estimate, with high accuracy, the number of physicians, cooks, or mechanics, and the quantities of every kind of material, reaching into thousands of categories. For example, in the American operation in Santo Domingo in 1965 it was possible to know in advance that the ultimate troop level would be 120,000 men, and to know the sizes, types, and rates of flow with which troop

mainchance radicalism in the first place, and in particular the aim of minimax to minimize one's own losses. When information analysis leaves no doubt that a completely devastating enemy strike is about to occur, a pre-emptive strike is called for to try to destroy completely the military capability of the enemy. This means it must occur without any advance warning and it must accomplish its mission within the space of a few hours: a two- or three-hour war of complete destruction. Of course, everything depends upon the accuracy of the information-processing system that extrapolates the indicators of an inevitable devastation.

A specific case was presented by the Arab-Israeli war of 1967. Israeli intelligence and information processing may not have been completely computerized, but they were remarkably thorough. Israeli pilots knew in advance which of the Egyptian aircraft on the ground were real and which were dummies. They also knew the names and biographies of the Egyptian pilots opposing them. The relationship between information and weaponry was such that a pre-emptive Israeli strike could yield victory within the space of a few hours. This was the clearly stated basis of Israeli preparation, and, just before their strike, the Israelis became convinced by their information system that they were indeed marked for complete extermination. They reacted with the remorseless mainchance abandon of the condemned. They sent troops home on dummy leave permits to camouflage their true intentions. Then they struck. The conversion point had arrived: minimax was transformed into mainchance.

Happily, nuclear weapons were not available to either side, but the information processes and games theory are close to those applicable to bigger confrontations, so it is disturbing to hopes for human survival that the validity of the Israeli information extrapolation is precisely what the post-mortem experts are still debating. Was an Arab attack truly inevitable? Was the Israeli pre-emptive strike too limited in its scope? Was the result a "real" victory for the Israelis or is it reversible?

Everything depends upon the validity of the information processing system, for, as we have said, the logic of completely valid information is ultimately to convert minimax into mainchance.

How to Make War Inevitable

Computerized intelligence and war gaming can only grow and tighten their hold on human affairs, as techniques improve. The ultimate possibilities and risks are perhaps beyond our present capacity to forecast. But if we look at the most advanced ideas today we can at least see the likely trend of the next twenty years if military confrontations between nations continue. The place to look for these ideas is Vietnam.

The advances in information-processing technology may be the most significant contribution to warfare to come out of the Vietnam war. If, as is often claimed, Vietnam is the 1960s counterpart of the Spanish civil war, then computerized strategic information leading to pre-emptive strikes is the counterpart of the *Blitzkrieg* of World War II. The United States has been operating four intelligence analysis centers in Saigon, with computers reportedly operating twenty-four hours a day. Over a thousand workers add to their computerized files more than 100,000 items each month. Supplementation comes from the "Big Eye"—four converted Super Constellation aircraft constantly watching and reporting on everything from the air. Raw information is collected from every possible source.

The long-term aim is nothing less than to account for

all enemy material and troops. Ultimately the system will be able to apply to enemy forces the same procedures of accounting and analysis that computerized inventory systems in industry use to account for the amount and location of production items. Then field troops can be given accurate information about the size and location of enemy units and weapon emplacements. One operation was said to have been furnished with intelligence location maps whose plotted items were later found to have been eighty per cent accurate. In at least one case a prediction by the system of an enemy attack was ignored by the field commander with disastrous results. If the ultimate aim is realized it will produce such a high degree of accuracy that the predictions cannot be ignored.

For example, the system aims at compiling critical military information about each individual among the enemy ground forces. Then, day-to-day variations in the information about given individuals can serve as reliable indicators of enemy tactical intentions. It is not far-fetched to suppose that sufficiently accurate information would make it possible to project probable enemy maneuvers even before the enemy himself had made his decisions.

If military information processing ever produces truly reliable projections of probable enemy actions, the consequences are clear. In the first place, decision-making will leave the hands of both field commanders and political authorities to become an adjunct of the information-processing system; then the computers may indeed take charge. Secondly, the tendency will be to make all military operations pre-emptive.

It requires but little imagination—and available hardware—to project for the entire world the application of the advances in information processing about enemy activities that have been achieved in Vietnam. Already the

C.I.A. has much of the intelligence infrastructure that would be necessary. For example, the Central American Defence Council, established in 1964, gathers all Central American defence and intelligence ministers into a council whose function is to coordinate joint military actions in the event of communist aggression or insurgency. Each local army has assigned to it American military and intelligence advisers, in extension of the Vietnam pattern. With the world brought under continuous surveillance operations, capable of piping masses of strategic information into real-time computer-analysis systems, the military prospect is for the advent of an age of pre-emptive warfare, triggered and directed by computer.

To sum up: in the next two decades we must expect that the most significant contribution of computers to warfare will be not the enhancement of weapons performance but the revolution in the handling of intelligence information. As we can already see, this revolution makes every incident a crisis and pushes every decision to the highest levels of responsibility: it puts political leaders in charge of military affairs but makes them more likely to use force. The time for decision in response to enemy action has shrunk from days or hours to seconds and will become in a sense negative, when future intentions of the enemy can be predicted with reliability. Then the computer will be all-important, and men will have to decide whether to believe what it says, because its characteristic strategy will be the pre-emptive strike.

MILITARIZED OCEANS

B Y

WILLIAM A. NIERENBERG
United States

The author is director of the Scripps Institution of Oceanography and professor of oceanography in the University of California, San Diego; he was formerly professor of physics at the Lawrence Radiation Laboratory, Berkeley. Professor Nierenberg was concerned with naval aspects of ocean science while director of the Hudson Laboratories, Columbia University. He served a term as Assistant Secretary General of NATO for Scientific Affairs.

THE TOTAL POTENTIAL for war in the future will be largely determined by its undersea component. The salient factors are the submersible vehicle and the commercial exploitation of the oceans. Although it would seem that undersea warfare has been with us for a long time, it has as yet been only a term to distinguish military operations that took place slightly below the sea's surface from those taking place slightly above the surface.

Until very recently the submarine was hardly an undersea vehicle at all. With its dependence on air for its diesels and with electric batteries that were technologically outmoded almost when they were introduced, it was a vessel extremely limited in speed (slower while submerged than any surface ship) and noisy and easily detected while on the surface. Before the advent of the nuclear submarine, the operational submarine spent only a small fraction of its time under the surface.

The nuclear submarine was the first step toward the true submersible. A vessel came into existence that, for all practical purposes, could remain submerged indefinitely and had unlimited range. It was capable of speeds to match any reasonable quarry. The Polaris system of intermediate-range ballistic missiles that could be fired from a submerged vessel gave the submarine a mission that switched it from a purely naval role to a full partnership in global warfare.

It is a slow history for reasons that remain obscure. The potential of undersea craft has been recognized from the earliest days, and long before the advent of nuclear propulsion, the fuel-cell concept was available that might have converted the clumsy pigboat of World War I into a real submersible; it was not used. This slow cadence of the past is rapidly giving way to an accelerando where war at sea may be the principal theme of global war.

The fundamental reason why the oceans are attractive to military planners is that, to a first approximation, the submerged weapon is invisible. Electromagnetic sensors of any practical use have limited range in sea water. The only reliable sensor of any range is acousticsonar, which detects submerged objects by their reflection or emission of acoustic waves. However, even today's relatively crude nuclear submarine employs a team well skilled in contending with sonar, and antisubmarine warfare has become an engineers' nightmare because of the need to exploit to the extreme this one sensor.

Ocean Science and Technology

In order to appreciate the future of undersea warfare, it is necessary to understand some aspects of the current state of the oceanographic sciences and their development. In no other field of warfare does the environment enter in so complex a way to fashion and fix the character of the operations. The classical example is that of biological intervention. In the air and on the ground the intervention of an animal or plant in military operations is rare and noteworthy. In the sea, freedom from such interference—by organisms that reflect sonar pulses, for example—would be extraordinary.

At present the science of oceanography is at the halfway point. It is clearly emerging with great rapidity from a purely exploratory first phase. On the other hand, whole regions are essentially unexplored and will open up entirely new possibilties. Some characteristics of the oceans are fairly well understood and subject to analysis and prediction; others still await identification and clarification. This unusual pattern of fundamental science and exploration is exciting to the working scientist; to the military, it is a challenge and a source of concern.

All of the fundamental discoveries in acoustics relevant to operational techniques, both for and against the submarine, have originated as by-products of oceanographic research and from the laboratories of universities and nonmilitary establishments. The effects on sonar of abrupt temperature changes between water layers, of the biological "deep scattering" layer, of the deep sound transmission channel, and of "internal waves" below the surface, are examples. The acoustic "appearance" of the sediments of the sea bed is coming to be understood more clearly, and their geographical variability will play a decisive role in the development and use of long-range sonar. On this basis, predictions of a military nature will depend a good deal on further exploration of the relatively unknown ocean basins. The complexities of the bottom and its general topographical features will offer still more security to the military submersible when it finally becomes fully operational at the greatest depths—far below those accessible to present-day submarines.

Until 1960 the importance of submarine warfare was almost solely related to the problem of interdiction of surface sea routes for the supply of expeditionary forces and allies. The concept of "control of the seas" was limited to this aspect alone and the question of "control of the depths" was never seriously considered. One reason for this was that strategic transport of men and material by submarine never seemed feasible; another, that absolute and universal interdiction of submarine activity seemed impossible on technological and political grounds. Finally, the sum total of human activity below the sea surface was minuscule. It is this last aspect that will change most dramatically in the next twenty years.

The last quarter of the twentieth century will witness the opening of a world-wide campaign for the utilization of resources on a global basis, including that major part of the planet covered by water. Future war at sea will depend on the rate of advance of two technologies exploiting recent and current ocean science: the military and the civilian. The signs are quite definite that the military technology will move ahead at the greatest possible rate. The civilian side is less clear.

Each of the technologically advanced countries has made

policy statements and taken specific actions both at the executive and parliamentary levels to strengthen its position in this presumably rapidly developing area of civilian ocean technology. The United Nations, through UNESCO and its other specialized agencies, is taking a very active role in trying to establish international cooperation in these affairs. Broadly speaking, these activities are grouped around fishing rights and development in local waters, supply of protein for global needs from world-wide fisheries, and extraction of mineral wealth from the continental shelves and from the bottom of the deep ocean. It is the last that may introduce the wholly novel civilian element in the future.

We are now, in the 1960s, just beginning to explore this vast but unknown marine province for mineral deposits. This decade of exploration will be followed by a period of development of the tools for the extraction of the wealth and finally by the actual installations for deep-sea mining. This great technological development will ultimately merge with the military technology. At that stage, many nations will have very large investments in the deep ocean, both on the bottom and throughout the ocean levels immediately above these installations. There will be teams of men working in these installations at all depths. We cannot foresee all the legal problems at this time, but we can tell that these massive investments will have a value that will make them attractive military targets, a source of international blackmail and friction, and, in general, a central concern of naval planners.

Of the same potential importance, but less clearly visualized at this time, are buoy fields and related ocean facilities for weather prediction and control. Other major civilian installations will undoubtedly be tapping the great reservoirs of thermal power represented by the difference between surface and bottom temperatures in the sea. This kind of installation will be very complex; it may include nuclear reactors to maximize the efficiency of power production and may yield by-products such as fresh water by desalinization of sea water.

Navies of the Future

Before reviewing some major aspects of future war at sea, we should examine the current position of the greatest naval power—the United States. It comprises a worldwide commitment of men and ships requiring an operating budget of the order of $20 billion a year. The U. S. Navy has 145 submarines, of which forty-one are of the Polaris type. It has 300 destroyers, twenty-four aircraft carriers, and perhaps 5,000 vessels of all kinds. In today's world it has proved capable of waging war at sea anywhere at any time, as demonstrated in Korea, Indochina, Lebanon, and the Dominican Republic, and during the Cuban crisis. Political developments, such as those surrounding the Panama and Suez canals, have led to major shifts in strategic deployment and in the development program. At the same time, the U. S. Navy has joined with the Army and Air Force in creating a credible nuclear deterrent.

The U. S. Navy achieved mastery during a period when other major naval powers declined. However, the U.S.S.R. has emerged as a great naval power second only to the United States. As a result, Russian vessels have been able to shadow the American fleet and seriously compromise its maneuvers. In the Israel-Arab crisis of 1967, the U.S.S.R. was able to exercise a certain diplomatic control over the U. S. Sixth Fleet in the Mediterranean. These complicated restrictions on American naval freedom will develop more rapidly as Soviet technology turns more and more seaward.

Russia has available, in numbers, a still larger submarine fleet whose intelligence and fleet-following abilities are very respectable. In addition, the Soviet Union has become a first-class maritime nation. The current return from its new world-wide fishing fleet makes its catch the second largest in the world. This is still more significant when we consider that the first, that of Peru, is dependent on a single crop from the phenomenal but nearby Peruvian current.

We can now look in some detail at the naval technology that will be developed by the 1980s. The chief trend in submarine technology will be toward operation at very great depths, and there will be some vehicles available to the fleet that will work to the bottom of the deep ocean. This ability will certainly be necessary for the hunter-killer submarine, which must be prepared to go anywhere to meet its objective. In general, the deep vehicle may not be just another submarine; it may be specialized equipment with or without men and with or without military significance. Its speed will not increase much, compared with that of present submarines, but its sensing and communications equipment will become more sophisticated. With civilian activities in progress under the sea, the assumption that any artifact detected by sonar is another submarine will no longer be valid; the sonars will have to possess discrimination of a high order. This submarine will be farther below the surface and its communication links will therefore be more highly developed. Despite its increased depth, its increased responsibilities for control of a three-dimensional region will force it to expose itself more, rather than to skulk behind natural acoustic screens. Hence it will require a more sophisticated set of defences against opposing weapons.

The most threatening naval vessel will remain the missile-carrying submarine. Technological advances will permit the dispersal of such fleets over all the oceans, and their missiles will be able to reach any point on the globe. Such dispersal combined with the greater operating depth will increase almost beyond measure the security and therefore the value of the system as a deterrent. However, the commander of a hostile submarine will no longer have the environment to himself. In peacetime he will have to share the oceans not only with well-instrumented fishing fleets and commercial vessels but also with undersea installations of all kinds—some of which will undoubtedly be pure "cover" for military posts. These installations, if numerous enough, could provide an excellent (perhaps the principal) source of reliable intelligence on hostile submarine operations.

The surface vessel is both a target and a threat for the submarine. Future solutions to the classical problem of keeping sea lanes open against hostile submarine forces will require considerable changes in technology at the surface. The increased speed of the submarine poses a far greater threat to the surface vessel and makes the submarine less vulnerable. This trend, combined with advanced surface-to-surface missiles, will stop the effective use of the sea for transport by conventional ships in time of war. The sea lanes would be closed, were it not for the introduction of high-speed "surface-effect" vessels, such as the hovercraft or the captured-air-bubble vessel (CAB).

These are machines that can operate in high seas at 100 knots (185 kilometers an hour) and can be built to 5000 tons and greater. This high speed, compared with those of the torpedo and the submarine, will reduce the loss of merchant shipping by torpedo attack to negligible proportions and put an obstacle in the way of the effective use of the surface-to-surface missile. Cargo-carrying costs will be greater than those of conventional ships in peacetime, but a large enough volume of cargo exists, requiring high-speed transport, that will help to justify the costs of a special fleet of surface-effect vehicles.

This kind of vessel is so promising that, by the 1980s, craft for amphibious landings will be using the surface effect. This change will affect the entire nature of the current assault force, which has only so much antisubmarine potential built into it, because of its low speed.

A use of the surface-effect vessel as a submarine hunter of a very deadly sort may be anticipated. Present difficulties of keeping up with a submarine contact in a surface ship can be largely overcome with a surface vessel that moves three to four times faster than the submarine. Some surface-effect vessels will be employed in this way because of the weight of armament and signal-processing equipment that they can carry. The burden of the hunt, however, will be carried by a new class of aircraft: one that can stay airborne for several days, make its contact quickly, and stay with it almost indefinitely. Both the surface-effect ship and this "oceanographic" aircraft will depend on widespread and sophisticated sensor fields that pierce the acoustic barriers inherent in the ocean structure. We must envisage the deployment, across the oceans and at various depths from the surface to the bottom, of fixed and mobile buoys and other structures that keep the surrounding waters under continuous sonar observation and register changes in the environmental conditions. Technologically, these devices will be similar to civilian buoys developed for ocean science and meteorology. They will be unmanned, but manned submersibles may be necessary to protect them from attack.

The aircraft carrier will evolve in a very complex way: it will acquire higher speed in the surface-effect mode and will have less need to carry antisubmarine fixed-wing aircraft. Its development will require compromises of choice.

While the surface-effect submarine hunter and the ocean-ographic airplane will tend to reduce the need for the aircraft carrier in its characteristic antisubmarine role, a carrier that can operate at 100 knots will offer a much wider choice of fixed-wing aircraft for operations. But this possibility will be challenged by the possible introduction of vertical-take-off aircraft, operating from fixed bases. The problem is then related to that of maintaining bases in other countries, which has several different diplomatic facets.

The ultimate solution may be the construction of giant floating bases stationed against wind and current in a global array that would protect strategic sites at less long-term cost than the current bases or carriers. Although this is a complex subject, much can already be said about it because of the conceptual relationship of such bases to the current, rapidly developing technology in offshore oil exploration and production. We can possibly diminish an important international irritant this way; but, as we shall see later, we may be replacing it with a more difficult problem.

The most important characteristic of submarine operations is their clandestine nature. To reinforce it, their base installations will also undergo major changes. They will, in fact, no longer be on shore but built into the edge of the continental shelf or its slope and linked to the shore by a variety of communications and transport facilities. These installations out at sea, on the continental margin, will provide the needed security. Normal logistics, including crew changes, will take place away from shore; equipment check-points, communications centers, and command posts will be as effectively hidden as the missile submarine itself. The military command and control system for undersea operations will depend on satellite communications with the mobile elements as well as with the ocean-wide fields of sensory devices. Information will be continuously acquired about the complete environmental state of the ocean as well as about the man-made vessels and installations it contains, both on and beneath the surface.

In the 1980s it will no longer be possible to speak meaningfully of antisubmarine warfare or even of undersea warfare. The development of new carriers and other surface vessels, and the better technological adaptation to the environment from the bottom of the ocean to the surface and the air immediately above, will lead to a new concept of total naval warfare in which all the elements are engaged at once. From a narrow point of view, we can sum up by saying that, in this concept for the future, the submarine and its opponent have maintained an even match. A closer look would indicate that while the submarine has lost tactically, because of the new technology, it has gained strategically. However, this viewpoint is indeed too narrow. What transpires in the concept is that the various elements of naval warfare fuse into one global war at sea in which all the elements are brought into play to greater or lesser degree depending on the nature of the confrontation. We have a vast "no man's land" with fleets moving at will through each other and each other's installations. Major effort will be expended to obtain up-to-date and meaningful intelligence because of the shadowy nature of these operations.

Instabilities and Rivalries

Even with the best intelligence, the system as foreseen has grave instabilities. A limited war in so complex a system can raise naval activity to the point of overtly endangering the security of a deterrent missile system. The gravity of this kind of accidental result has no simple appeasement.

If we look beyond the 1980s, we can see the civilian undersea technology being rapidly dispersed among the smaller countries and the costs falling to a level where small companies can operate effectively and in large numbers. Since national boundaries are no longer involved, the security of a strategic system would become greatly impaired. As a result, we may see a startling reversal of policy on the part of the great powers. Huge expanses of the oceans may be seized, de facto, and isolated for their own strategic purposes, thus limiting innocent passage on the surface. This is a serious consequence of the technology, and very few alternatives are available, other than the total abandonment of the oceans as a strategic preserve. The construction of giant floating air bases, mentioned earlier, would entail problems of protection that would certainly encourage such a policy of excluding foreign vessels from a large volume of the surrounding ocean, in defiance of traditional rights of passage.

Naval warfare will develop more incoherently than other aspects of warfare because of the lack of national boundaries to contain activities in a traditional way. Apart from fishing, the role of the oceans in the past was that of a transportation system for men and weapons, for commerce and for the means of interdicting commerce. The development of the submarine strategic system and the possibility of exploiting the ocean basins for material purposes have added possession of the oceans as an objective, even if in a limited sense of the word. The drive toward such possession, and the accompanying technology, will return naval warfare to its earliest origin, which was that of economic domination by the country that had most succeeded. The cost will be immense. Even by the standards set by the space program, the initial development costs of deep-submergence technology will be considered high.

This conclusion is extreme and is, of course, debatable. The controlling forces in the evolution are perceptible, but their effect is conjectural. The law of the sea as applied to the deep ocean (beyond the continental shelf) is, at best, discussed by analogy with prospectors' rights that have developed on land. The concept of contiguity is important in this connection; namely, that the discoverer of a body of ore has certain rights in the neighborhood of his find. Applied to the deep oceans, this concept would have to be developed more fully than seems reasonable. Initially, few countries will have the technical ability to exploit a discovery, and strong political forces will develop behind the commercial and military deep-sea technologies. The demands of these forces can be heard today.

There is a well expressed idea for turning over the deep-sea domain to the United Nations as the ultimate sovereign, with the licensing fees from development going to that organization. This proposal is worth examining, but from the viewpoint of the stability of existing deterrent systems, it has implications that the proponents have either ignored or felt not worth considering. On the other hand, there is a segment of government that believes that the first countries to develop deep-submergence technology will be the first in strategic possession. It is not evident that the proposers of such a "tough" policy have clearly in mind the international or strategic implications of their ideas, any more than those who trust to U.N. ownership. Whatever the outcome, the nations of the world will very soon have to make important decisions with regard to this seventy per cent of the earth's surface; and they must do so from an entirely different point of view from that which governs the thirty per cent of dry land.

THE TOXIC ARSENAL

B Y

MARCEL FETIZON

AND

MICHEL MAGAT

France

Both authors are in the Faculté des Sciences, Orsay. M. Fetizon is professor of thermodynamics, and his research is particularly concerned with the synthesis and mass spectrometry of natural products. M. Magat is professor of physical chemistry, working principally on the chemical effects of ionizing radiation, high polymers, and molecular crystals. During World War II, he worked for the Free French, in operational research for the R.A.F. Fighter Command. He is now a participant in the Pugwash movement.

IN ORDER TO MAKE a calculated guess concerning the chemical weapons that will be available in the 1980s it seems useful to trace a brief history of these weapons. In this way one can hope to find the internal logic of their development and to uncover the underlying trends. Indeed, medium-range prediction is possible only as a logical extrapolation of what already exists or is under study. Unexpected discoveries—which may be the most important ones because they may revolutionize the situation—are in essence unpredictable except in very long-range extrapolations wherein one has to rely more on dreams and intuition than on scientific logic.

Chemical warfare is probably very old, nearly as old as the human species. What has changed is the scope and the techniques of delivery, which in turn influence the nature and the physical form of the products used. Indeed, so-called "savages" used, and are still using, poisoned arrows. The natural agents employed are curare and other vegetable poisons acting primarily on the nervous system. The limitations of these weapons and poisons are manifold: the supply of poisonous substances is limited, penetration into the body requires a wound, and only one person can be killed by each hit.

The use of incendiaries is more recent, since it requires in general a better knowledge of chemical properties. By 429 B.C. the Spartans were using a mixture of pitch and sulfur to set a city on fire; American Indians had incendiary arrows, and "liquid fire" for military purposes figures in Assyrian bas-reliefs. By A.D. 350 the Greeks had added naphtha or petroleum to the inflammable mixtures, and during the seventh century Callinicus developed "Greek fire" in a primitive version of flame-throwers—burning liquid thrown out from siphons—which remained in service until the Middle Ages. Smoke screens served to cover troop movements and assisted, for instance, the Swedes in their crossing of the Dvina river in 1700.

It is remarkable that poison was not used in war by "civilized" societies, or even thought of until 1862. During the American Civil War a proposal was made to use chlorine-filled shells, but the proposal was rejected by the United

States Government. It is difficult to know exactly what the reasons were for this lack of interest. Probably the progress of physical methods of warfare related to the development of explosives was partly responsible for it, perhaps also the inability to produce and to deliver large enough amounts of the poisonous substances known at that time. Ethical considerations may also have played their part, poison being more or less synonymous with treachery.

The progress made by the chemical industry during the nineteenth century lessened some of the technical objections. And, faced with the dangers of chemical warfare, the Hague International Peace Conference in 1899 adopted a resolution "to abstain from the use of all projectiles the object of which is the diffusion of asphyxiating or deleterious gases," on the ground that this method of warfare was inhumane. Among the big powers, only the United States did not support this resolution—on the grounds that the inhumane aspects of this type of warfare were not clearly established. (This seems to have been a consistent viewpoint of the United States Government, except in the early 1920s, and in 1926 the Geneva Protocol repeating the ban on chemical weapons was buried in the Senate because of opposition by the American Legion and the American Chemical Society. Despite the declarations of Roosevelt [1943], Eisenhower [1960], and others, "riot-control" gas has been used in Vietnam together with "defoliants," the Defense Department stating that, due to recent research results, "some forms of the weapons . . . could be effectively used for defence purposes with minimum collateral consequences.")

During World War I gases were used for the first time on a large scale. Although the first use was probably the attack by the French in 1914 using a "riot-control" tear gas, the "official" date for the beginning of gas warfare is 22 April

1915, when the Germans launched an attack by chlorine. At that time both sides were firmly entrenched and the smallest advance cost a tremendous number of lives. Gas was launched by opening cylinders and relying on the wind to carry it toward the enemy. An unexpected change in the wind's direction could carry the gas toward the attacking troops, which seems to have happened on another occasion. We can see immediately the "progress" compared with poisoned arrows. The toxic substances do not penetrate in the blood stream through a wound, but are absorbed by inhalation and attack the eyes and the respiratory tract. The poison acted not on an individual (one hit—one casualty) but on groups.

The means of delivery was nevertheless deficient, since it could be done only at short range, with the danger of a boomerang effect. Defence was relatively easy; very crude gas masks were efficient. The effect of the gas attack was immediate but not persistent, because the gas was dispersed by wind. All efforts were made on both sides during World War I to alleviate these defects: bomb shells replaced gas cylinders, allowing an attack from longer distances. They then developed longer-lasting chemical weapons: mustard gas (bis 2-chloro-ethyl-sulphide), nitrogen mustard [bis (beta-chloro-ethyl) amine hydrochloride], and lewisite [dichloro-(2-chlorovinyl)-arsine], slowly evaporating liquids insuring a persistency of several hours. Phosgene, although not persistent, had immediate deadly effects. The possibility of attack with several separate products or their mixtures made protection by the gas mask more complicated. The casualties were more numerous and more severe. It is true that the attack by mustard-gas shells in 1918 in the Ypres–St.-Quentin area caused only eighty-seven immediate deaths among the 7000 casualties, but the number of people disabled for life and dying later of the consequences of mustard-gas inhalation was many times larger.

During World War I chemical weapons were used only against troops and never against civilians. The first to use mustard gas indiscriminately, by air bombs and aircraft spray, seem to have been the Italians during the Abyssinian war. The casualties among the unprotected Abyssinians are said to have been terrible. As effective—and for the same reasons—were the chemical attacks by the Japanese against the Chinese.

World War II was characterized by an absence of gas warfare, although (or perhaps because) both sides were prepared for it. Chemical warfare was limited to incendiaries and flame throwers. Incendiary bombs with white phosphorus were used on a large scale by both sides against cities. When exploded by the bursting charge, solid phosphorus disperses in the form of small particles that ignite by contact with the air. Ignited particles cause painful flesh burns that heal very slowly. However, phosphorus bombs had only a limited value and were superseded by metal incendiaries made out of magnesium and powdered aluminum mixed with finely dispersed zinc and iron oxides. The most efficient incendiary, usable both in fire bombs and in flame throwers, proved to be napalm—gasoline thickened by aluminum soap of naphthalenic and palmitic acids. It was used on a large scale by the Allied forces against cities and also against troops, particularly troops in trenches or bunkers. After World War II napalm was used by the United States in Korea and in Vietnam, and by the French in Algeria. It has a devastating effect on enemy forces, men being transformed into nearly inextinguishable living torches. Napalm seems to give entire satisfaction to the military and will probably still be used in the 1980s, perhaps with minor improvements.

Research on toxic gases accelerated during World War II

even though they were not used. In Germany a very important technical advance was made: the discovery of a family of three highly lethal "nerve gases" known by the names of Tabun, Sarin, and Soman. All three are derivatives of phosphine oxide. They penetrate either through the respiratory tract or through the skin if deposited as a small drop, without producing blisters or itching. The latter method of entry is, of course, of paramount importance because protection has to be extended not only to the respiratory system but to the whole body surface. Since some of these gases have no odor and since they produce no skin irritation, no warning is given to the men affected before it is too late. Clearly, protection even of military personnel against nerve gas is very difficult and virtually impossible as far as civilians are concerned. Since 1945 some new substances having the same mode of action—organophosphorus compounds such as dialkylaminoalkylthiophosphonic acid, and metal carbonyls—have been added to the list of nerve gases.

The effects of nerve gases appear immediately or in a span of fifteen minutes. They are consequences of the inhibition of cholinesterase, an enzyme essential to the proper control of muscles by nerves. Since gases of this type may well still be in use in the 1980s, we shall describe their effects in more detail. According to U.S. Army Technical Manual TM 3-215, there appear in order: running nose; tightness of chest; dimness of vision and pinpointing of eye pupils; difficulty in breathing; drooling and excessive sweating; nausea, vomiting, cramps, and involuntary defecation and urination; twitching, jerking, and staggering; headache, confusion, drowsiness, coma, convulsion, cessation of breathing, and death. Skin absorption great enough to cause death may occur in one or two minutes; death may be delayed for one or two hours. Respiratory lethal doses and liquid in the eye kill in one to ten minutes.

The lethal dose of Sarin is about thirty times smaller than for phosgene, considered as the most efficient lethal gas before World War II. This means that the lethal dose is 0.01 milligrams per kilogram of body weight, which corresponds to about 0.7 mg. for an adult and 0.1-0.3 mg. for a child. The dose for an adult can be absorbed in a few minutes if the concentration of Sarin is of 0.1-0.3 mg. per liter of air (in one inhalation if the concentration in air reaches 2-4 mg/l). On this basis, it is easy to see that in order to reach a lethal concentration in the air of a city the size of Paris, to a height of fifteen meters, some 250 tons of Sarin have to be distributed. This is a very small amount, considering that the German stockpile of Sarin in 1945 was of 7200 tons and that 250 tons can be carried nowadays by some twenty to twenty-five bombers. By the 1980s, this will probably require about the same number of missiles.

Besides the post-1945 additions to the nerve gases, a large effort has been made in the direction of psychic poisons, to be discussed later. Highly effective plant poisons have also been developed—such as the defoliants used by the United States in Vietnam to kill all vegetation and crops over a target area. It is well known that all the major powers, at least, are continuing and increasing research in the field of chemical warfare. Although this kind of research is done openly in the United States (where figures on expenditure and research manpower are made public), it is more secret in other countries. For instance, the existence in West Germany of a very large and powerful, highly secret laboratory working on problems of chemical warfare is known only from rumors.

It is obvious that the results of such research are highly secret and can be kept secret much more easily than can developments of nuclear weapons or missiles because experiments can be conducted in a much less spectacular way. It is impossible to keep nuclear explosions secret and very difficult to do so for missile tests. But one can experiment with chemical weapons with only very few people being involved. This has two consequences: first, that all countries doing nuclear or space experiments are publicizing what they can do, retaining only the "know-how"; secondly, that the public at large is sensitized to nuclear weapons but inclined to overlook the dangers of chemical warfare.

It is interesting to note that it is often claimed concerning other types of weapons—nuclear bombs, long-range missiles, and so on—that civilian production ultimately profits from experience gained during the development of military applications (civilian "fallout"). This contention seems to be true to a certain extent for some of the weapons, e.g., missiles, although the question can be asked whether the same benefits could not be obtained at a smaller expense by investigating purely civilian applications. However, this contention is definitely not true as far as most of the recent progress in chemical warfare is concerned. Psychic drugs have been developed in the first instance for medical purposes, and plant poisons proceed from the agricultural research aimed at weed-killers. It seems that on these occasions it is civilian research that "feeds" the research for military applications. This of course does not apply only to the latest developments. The industrial production of chlorine and phosgene is not owing to their military applications; on the contrary, their applications for war were a consequence of the existence of these materials on the industrial market.

The Choice of Agents

Let us briefly summarize the arsenal of chemical weapons already available. We shall classify them according to their primary target and indicate their possible future utility.

Against buildings, the available chemical weapons are incendiaries: phosphorus and oxidizable metal powders. These are powerful weapons against countries where there is a high density of buildings made of combustible material or having at least a large proportion of combustible components. They lose their importance when most of the buildings are in concrete with metal frames. Since in most developed countries (except perhaps the United States) the general tendency is in favor of the latter type of architecture, it is very probable that by the 1980s the chemical incendiaries will play only a minor role in conflicts involving highly industrialized countries.

A second type of target is people in bunkers and the like. The available weapon, napalm, can be used against both armies in the field and civilians, and, as already mentioned, napalm will probably be so used well into the future.

Thirdly, there are military and civilian personnel in general, against whom toxic gases are effective weapons. The toxic gases can be subdivided in several categories according to the severity of injuries and possibility of protection. The least offensive are tear gases like chloracetophenone. The effect is immediate, but the gas is easily dissipated. Defence by gas mask is generally available in time of war. This is a typical riot-control gas, useful only against unprepared civilians. Much the same can be said concerning vomiting gases of the adamsite type: the consequences of inhalation are not very severe except when administered to children or undernourished adults.

A more potent group is formed by so-called "blister gases" (distilled mustard, nitrogen mustard, lewisite, and the like). These gases produce blisters and inflammation of respiratory tracts, eventually developing into bronchopneumonia, diarrhea, etc. They may lead to death or to permanent incapacitation even of healthy adults. Protection by gas masks is quite efficient, but the blister gases may still be used, since the carrying of gas masks hinders troop mobility. This type of gas will probably not be employed in a war between technologically advanced nations in the 1980s.

The next three types of gases are lethal even for healthy adults.

The "blood gases": hydrogen cyanide, hydrogen chloride, and arsines. The first two interfere with utilization of oxygen by hemoglobin in the blood and also poison the central nerve system. The action is in general immediate. They proved their efficiency in confined spaces: under the name of Zyklon B, hydrogen cyanide and its analogues were used by the Germans in gas chambers of extermination camps. Arsine interferes with the functioning of blood and attacks the kidneys and liver, but its effects are delayed by several hours or even days. It is hence practically useless on the battlefield proper. Protection by gas mask against these gases is efficient, and it seems probable that they will not be used in a major war in the 1980s, although their use against guerrillas, particularly when mopping up dugouts, may be envisaged.

The "choking gases": phosgene and diphosgene. These act on lungs, causing swelling that deprives the victim of oxygen. The effect is delayed by three hours or more. The gas mask is efficient, and the gases are not persistent. They will probably not be used by the 1980s.

The "nerve gases," which we have already described in some detail. They will probably rank high among the gas weapons of the 1980s.

The big advantage of gases as compared with explosives, from the military point of view, is that they attack enemy forces, military and civilian, without destroying buildings or equipment, thereby abolishing the material problems of postwar reconstruction that gave rise to such difficulties after World War II. This consideration would be even more important in wars of conquest waged to obtain new territory for resettling excess population.

Finally, a fourth target category is crops and vegetation. The goal may be strategic (destroying crops and trying to create famine for the enemy) or tactical (destroying leaf cover used by the enemy for concealment). A fair number of materials related to plant hormones have been discovered for civilian weed control. Examples are 2, 4-dichlorophenoxyacetic acid, killing herbaceous plants, and 2, 4, 5-dichlorophenoxyacetic acid, effective against woody plants. These compounds are highly efficient: about a kilogram is enough to "weed out" a plot 200 meters square. A doubled dose is toxic also for cultivated plants. Even more efficient is dimethylarsenous acid: doses of such herbicides for overall destruction of plant life are of the same order as those of mustard gas for humans.

Such materials are potentially very effective against small countries or isolated guerrillas drawing their sustenance from relatively limited areas. It is doubtful whether they can be used in major conflicts involving countries with large cultivated tracts because they must be fairly evenly distributed over a very large area, and that can be done efficiently only by relatively low-flying aircraft, which presuppose mastery of the skies. Once such mastery is achieved in conventional warfare, the final victory is probably not far off, in any case. Even in the future defoliants and plant killers will be used only against guerrillas or in other revolutionary wars.

So far as developed countries are concerned, much more effective would be biological weapons (diseases of the crops) or the burning of the crops just before harvest by high-altitude explosions of H-bombs. Another possibility, as yet untested, would be the creation of a "hole" in the ozone layer of the atmosphere above enemy territory (see also Professor MacDonald's chapter). It is well known that the existence of this ozone layer, absorbing deadly ultra-violet radiation from the sun, is a prerequisite for the existence of life on land, since the spectral region concerned is absorbed by many organic substances, among them proteins, and leads to their decomposition. Since ozone itself reacts very readily with many organic compounds, it is possible that at least a partial destruction of ozone in the region where it is most abundant (twenty to forty kilometers above sea level) would be achieved by dispersing a convenient reagent at this height. It is not unbelievable that this could be achieved by the 1980s.

Psychic Poisons

Let us now examine the latest additives to the arsenal of chemical warfare, the so-called psychic poisons, or psychotomimetic compounds, often referred to in the military context as "incaps" (incapacitating chemicals). They are intended to act on the brain and produce "temporary" mental disorder among the opponent's military and civilian personnel.

Drugs exerting such action on the brain have been known for thousands of years in many countries as crude extracts from plants and mushrooms. It seems that these substances are widespread all over the world, but that they are particularly abundant in Mexico. In 1888 Lewin discovered that the chewing of "mescal buttons" or "peyotl" (dried tops of a cactus growing near to the Mexico-United States border) had long been a practice among Indians and produced strong hallucinations. The synthesis of mescalin, the compound responsible for the pharmacological activity of peyotl, was achieved in 1919. Mass production of mescalin would not be a problem for the chemical industry nowadays, but the necessary dose is too high for military applications. Other plants or mushrooms were, and still are, used by Indians of Mexico, Guatemala, and Brazil, by Negroes in Haiti, by natives of New Guinea, and also elsewhere, in religious ceremonies in order to produce ecstatic states. However, here again, the necessary amounts are relatively large.

An entirely different set of psychotomimetic drugs has been found in animals and plants. These drugs are characterized by molecules containing a so-called indole nucleus, also present in very useful substances such as the hormone serotonin, which seems to be required for the human brain to work properly. For instance, bufotenin, isolated from the skin of many toad species, when injected into a dog makes it apathetic and incapable of defence. Curiously enough, the same compound has been found in seeds of certain Leguminosae used in the West Indies and South America by sorcerers for various religious purposes. Psilocybin and psilocin, closely related to bufotenin, are the active compounds of some Mexican mushrooms and are used collectively under the name of teonanacatl ("God's flesh") by the Indians in their ceremonies. "Ayahausca" and "yagé" are also used by South American Indians as psychotomimetic drugs, for the production of ritual hallucinations. The plant species responsible for this activity have been studied, and various materials have been found, all chemically related to bufotenin. Psilocybin was first isolated in 1958 and its structure and synthesis described in 1959. An oral dose of a few milligrams is capable of producing the same alteration in human behavior as bufotenin, but no injection is required. Here again, production of psilocybin on a large scale is now possible.

Many other compounds related to these indole alkaloids, but not found in nature, have been synthesized; but in general, as far as is known, their biological activity is rather low. There is, however, no reason why a synthetic substance should not be much more potent than its natural counterpart: examples are known, for instance, in the steroid series of hormones, where compounds prepared in the laboratory and not found in living cells are several hundred times more effective than, let us say, cortisone. Nature has not necessarily found the best chemical to produce a given activity. It is possible that synthetic compounds far more potent than bufotenin or psilocybin have been discovered, or soon will be.

The root of an African shrub, *Thabernanthe iboga*, contains a material called ibogain, whose structure is far more complex than the aforementioned drugs and which shows potent hallucinating properties. A great deal of work has been devoted to the synthesis of this compound, which has recently been described.

By far the most potent of the disclosed psychic poisons is the diethylamide of lysergic acid, also known as LSD 25, or simply LSD. Lysergic acid occurs in chemically combined form in ergot, but total artificial synthesis of lysergic acid was achieved in 1954.

Up to now, a great variety of lysergic acid derivatives has been tested, but it seems that the most efficient is still LSD, which is able to induce behavioral changes in man at the dose of less than a microgram per kilogram of body weight. This means that one tenth of a milligram taken orally is enough to cause optical hallucinations and even mental states resembling schizophrenia. In other words, a kilogram or so of LSD is, in principle, sufficient to render temporarily schizophrenic the entire population of London, in the ideal case of even distribution. But even assuming that only a thousandth of the LSD distributed is taken in by the population, the quantity necessary is only one ton. Although the most efficient method of distribution would be the poisoning of drinking water, inhalation of fine dust could also prove quite efficient. The whole population of a country could be poisoned by spraying LSD solutions over large areas, which seems technically possible today.

During a medical conference at SHAPE (the Supreme Headquarters Allied Powers, Europe), a newsreel was shown that gave an idea of the behavior of a battalion of soldiers "treated" by LSD added to their morning coffee. The soldiers were laughing without reason, throwing their guns away, climbing everywhere, screaming, weeping just like children—behavior far from normal in an army. Gay war, funny war, humane war . . . ? At first glance, perhaps. Looked at more closely, it is not as funny or humane as it seems.

The schizophrenic state induced by LSD seems to be reversible if the drug is taken under the control of a physician, provided the dose is low enough. But grave, not completely reversible, effects have been observed with LSD users in cases of overdoses. Used in a war, where it is intended that everybody should get an effective dose, most of the population will have to receive a large overdose. Such an overdose may either drive the victim mad for the rest of his life or simply kill him. And what is an effective dose for a man may be deadly for a child or for a pregnant woman. In addition it was recently announced that LSD possesses teratogenic, i.e., monster-producing, properties.

Again, many people receiving LSD will not be sitting quietly at home. They will be driving cars, trucks, and trains, directing traffic, handling deadly weapons, giving orders. And animals will go wild. Reactions to LSD are unpredictable, and it is hardly likely that there will be no loss of life in the ensuing chaos.

Other classes of compounds such as atropine and tropane derivatives, and the closely related piperidyl esters, such as "ditran" (1-ethyl 3-piperidyl cyclopentylphenylglycolate), have also been investigated as potentially psychotomimetic drugs. It is obviously very difficult to obtain accurate data for these drugs because of military secrecy.

All the compounds mentioned so far are alkaloids or derivatives therefrom, that is to say, nitrogen-containing substances. However, there is also a class of nitrogen-free chemicals, which shows a rather potent activity: the cannabinols. Under the names of marijuana (the Americas), hashish (Middle East), or kif (North Africa), cannabinol-containing drugs have long been recognized as able to induce a feeling of well-being, distortion of space, and double consciousness. The active dose is far greater than for LSD, but progress can be made and present knowledge enormously widens the field of potential "incaps."

All these drugs act like LSD, causing to some extent a state of temporary schizophrenia. The details of their action are not yet firmly established. They are supposed to interfere with normal metabolism of 5-hydroxytryptophane, a precursor of serotonin. But it is also suggested that they could have an effect on adrenalin in the body, and there is some evidence that some psychotomimetic substances may well act by modifying the balance of the essential material, acetylcholine, in the brain. But whatever these biochemical explanations may be, the fact remains that it is now possible to poison a whole country, creating a psychotic state simply by spraying psychotomimetic compounds. It can be expected that by the 1980s more "progress" will be made.

In the United States the incapacitant now standardized for military use is known under the name of BZ, its precise nature being kept secret. U.S. Army Technical Manual TM 3-215 lists the effects of BZ intake as: "interference with ordinary activity; dry, flushed skin; tachycardia; urinary retention; constipation; slowing of physical and mental activity; headache; giddiness, disorientation, hallucinations; drowsiness and sometimes maniacal behavior." There is also a warning: "critical limitations" do exist for the use of BZ, although it is considered useful against intermingled enemy and friendly military units.

It is probable that in a war around 1984, psychic poisons will be used particularly, since the necessary amounts can be prepared by a small number of skilled chemists and since their delivery can be made by saboteurs. And these, as we have seen, are only one component of the chemical arsenal.

Chemical weapons, if less spectacular than nuclear bombs and usually attracting less attention, are also efficient weapons of mass destruction. As such, they should be banned and controlled, although their control will doubtless pose very difficult inspection problems. But the survival of humanity depends as much upon this as upon anything else.

Considering the arsenal of chemical weapons already available and the degree of "perfection" attained, one might imagine that no further "progress" will take place. Unfortunately, this is too optimistic an outlook. Military research never stops but tries to produce more and more terrible weapons. In the type of arms we are considering here, attempts will be made to find more potent compounds in order to reduce the active dose and to make the delivery as easy as possible. The "progress" and its rate will depend essentially on civilian progress in two directions: chemistry of natural products and chemistry of life and mental processes. In organic synthesis we have reached the point where, when an active substance is well identified and its structure established, it is possible (if need be) to develop an industrial process for its production by a very few years' work. Manufacture will definitely not be a bottle-neck.

In the field of study of natural products, while the study of poisons produced by plants is already well advanced, the systematic study of poisons produced by animals is only just beginning. For instance, there is a kind of fish, well known in Japan as Tora Fugu, that produces a highly toxic principle, soluble in water: tetrodotoxin. Its structure is now established. The lethal dose could be as low as 0.5 mg. for an average-sized man. Many other powerful poisons are probably made by invertebrates and fishes, and by marine plants. Very little has been investigated so far, in these areas.

So far as biochemistry is concerned, it is to be remembered that many of the complex processes within the human body depend on very small traces of vital materials and consist in general of long sequences of reactions that are essential for continuation of life or for proper functioning of the mind. A perturbation in any one of the steps of such sequences can lead to death or mental disorder. Such interference is (at least in principle) possible at many stages, particularly when enzymes are involved; a typical example are the "nerve gases" we considered above, which interfere with acetylcholine esterase. It is not impossible that substances will be found of which the toxic dose will be of the order of a few micrograms.

The further we go in our understanding of biochemical

processes, the more possibilities for interference will develop. Modern medicine and biology are now making big strides in disentangling these life processes, and we have already learned how to interfere with many of them: immune response (protection against infection), cell growth, cell division, etc. The discovery of "messengers" regulating the production of specific enzymes indispensable for life processes gives new points of attack; so does the discovery of allostery (conformational changes of enzyme molecules induced by chemicals, which can enhance or reduce their activity).

Psychopharmacology is a new branch of science, less than twenty years old in its modern form. But we already know how to induce or attenuate at will certain mental disorders like schizophrenia. We have good reason to believe that great progress will be made in this field during the coming years, although the course of this progress is unpredictable. It is hence at least probable that more subtle and potent drugs than LSD, able to control mental processes and mood, will be discovered.

It is a terrible commentary on human nature that each piece of medical or psychopharmacological research undertaken for humane purposes in order to alleviate illness and save life, and each fundamental biochemical investigation undertaken in order to understand and satisfy our mind, may lead to possible military applications.

The question may arise: Is all science damned? We must either eliminate science or eliminate war. We cannot have both.

THE INFECTIOUS
DUST CLOUD

BY

CARL-GÖRAN HEDÉN
Sweden

Professor Hedén is a microbiologist working at the Karolinska Institutet in Stockholm, distinguished for his researches in microbial physiology and bioengineering. He is a member of the Medical Research Council of Sweden, and has played a prominent part in the Pugwash study of the possible control of biological weapons, to which he refers.

THE DISCOVERY of new weapons often becomes the subject of sensationalism and exaggeration, and one tends to forget that there is normally a considerable time lag between laboratory results and operational "hardware." However, this interval may be smaller in the case of biological weapons (BW) than for many other types of armament. There are now enough indications in military innovations in the biological field to predict a weapons system that will offer an enormous offensive potential at the same time as it will introduce staggering defence problems.

The offensive potential of biological weapons depends on five main factors:

1. the possibility of choosing a microorganism or a toxin tailored to the military need, whether it is to incapacitate the opponent temporarily (e.g., with certain viruses), to eliminate him permanently (e.g., with plague), or to attack his crops (e.g., with wheat rust);

2. a high effect on the human, animal, or plant targets, in relation to the weight and cost of the weapon;

3. a psychological effect based on the likelihood that very large groups of civilian and military personnel would be affected;

4. the imperceptible and insidious character of the attack, involving extremely difficult defence problems because of the large number of methods of delivery potentially available to the enemy;

5. the absence of physical damage to buildings and other structures would enable the attacker to take over the material resources of the territory.

Although infectious diseases have always been a serious problem for armies, biological warfare has, with few exceptions, been attempted only in a crude way and never as a major weapons system. In 1942 a military microbiologist could express the view that "it is highly questionable if biological agents are suitable for warfare" and as late as 1958 the American civil defence authorities still regarded the risk of biological and chemical attack as slight. Since then, however, there has been a re-evaluation of the situation, and several special investigations have emphasized the advances made in the handling of biological aerosols (dust clouds) and have put biological and chemical agents in the same class as nuclear weapons. One now comes across authoritative statements like: "The offensive use of biological agents is feasible," and "Biological agents exist which can be used strategically to cause casualties in an area the width of a continent." A Russian colonel in 1959 went as far as to say that "from results of comparative studies of the losses of life from conventional weapons, war poisons, and atomic energy on one side and losses from biological weapons on the other, it is believed today that a biological war would have the greatest effect of all." Biological warfare has in fact given us the most forceful instance so far of the truth of Isador Rabi's observation that "the combining of military techniques and science makes it easy to apply scientific principles to kill people—who are not strong structures."

Very little has been published about the Soviet study of biological warfare, but a high level of civil defence preparedness indicates that it has attracted much attention; that would hardly have been the case were there no first-hand information on the offensive potential. In fact, it was claimed that, when the American program started to expand in 1959, the U.S.S.R. led the United States. The American level of expenditure for research and development in chemical and biological warfare then climbed from around $35 million per year to about $150 million by 1964, so it is conceivable that the supposed gap has now been closed, particularly if one adds the British and Canadian investments of funds and personnel.

Simple competition may partly explain the biological arms race, but technical factors have provided fuel for it. In the first place, strategy is becoming more and more dependent upon civil defence, and microbiological weapons are well suited for large-scale civilian targets. Some of their inherent characteristics, such as the incubation period and the possibilities for protection of one's own forces, add a certain amount of freedom to their strategic use and also offer advantages for counterinsurgency operations and in fighting limited wars.

Secondly, recent progress in medical and biological research is easily applicable to biological warfare. Our knowledge about the immunological and biochemical relations between the infectious agent and its victim has increased immensely. Methods for genetic manipulation of bacteria and viruses have been found, so that essentially new disease agents can be devised, against which defence preparations are almost impossible. Moreover, techniques for the mass production of most types of microorganism have been highly developed for the purposes of medical research and for making vaccines.

Again, earlier doubts about the striking power of isolated infectious agents have been rendered obsolete by extensive tests on animals and human volunteers. The doses needed for infection of humans have been established for several agents. For instance, about twenty-five inhaled cells of the bacterium *F. tularensis* are required to induce rabbit fever (a prostrating but not often fatal disease). In the case of Q-fever (a debilitating disease) even a single inhaled particle of *R. burneti* might be sufficient to cause infection, so that, in theory, three grams of embryonic chicken tissue inoculated with Q-fever might hold enough infectious doses for the entire human population of the world. Of course the logistics of distribution and the decay rate of infectivity are not considered in this estimate.

Finally, it has been shown that a large number of infectious agents and toxins can be disseminated in the form of aerosols. The discovery that diseases that are normally spread only by insects might be disseminated in this way is of particular interest because it means that in those cases there would be no person-to-person spread beyond the selected target area.

Present Inhibitions

Why, then, have biological weapons not yet been used on a large scale? This is a very difficult question to answer, but the following points may be considered:

1. that the military need has not yet arisen;

2. that critical problems in the production, storage, and delivery of some agents, judged necessary to make up a fully operational system, have not yet been solved;

3. that the potential attacker's own biological defences are not adequate or have not been fully tested. The existence of a sophisticated aerosol warning system in a country might for instance not be enough to induce that country to use infectious aerosols offensively, since this could invite biological sabotage acts that would circumvent that particular type of defence;

4. that the popular concept of biological weapons as terror agents makes their use most distasteful, at least in countries with a free press and a democratic system of government;

5. that the value of such weapons is relatively less to the nuclear-armed powers, which also have a great potential for biological warfare, than to the smaller, nonnuclear nations, which have a more limited capability;

6. that all the tactical and strategic implications—for example, the opportunities for small groups of men to bring about devastating reprisals—have not yet been given serious attention in discussions of military policy;

7. that the exact target area and the long-range ecological consequences are very difficult, and occasionally impossible, to predict. We actually lack much knowledge of the susceptibility of many of the mammals, birds, reptiles, amphibia, and insects that would be exposed during an aerosol attack. Even the most perfect physical protection of man

would be useless if he would later become the victim of newly established disease reservoirs in animals and insects;

8. that there exists a sort of *pactum turpae* based upon the unpredictability and complicated consequences of a biological attack. These must introduce very disturbing elements in the fashionable mathematical mode of military thinking. One simulated attack, for example, is said to have killed or incapacitated 600,000 friendly or neutral civilians at the same time as it eliminated seventy-five per cent of the opposing troops.

How long the period of grace may last, when it rests on as complex a foundation as this, is impossible to tell, but the psychological factors that underlie the Geneva Protocol of 1925 still seem to exert a strong inhibitory influence.

Delivering the Dose

The possibility of disseminating biological agents over very large areas has been demonstrated in field tests involving both inert particles and harmless bacterial spores. For instance, a cloud of inorganic particles (200 kilograms generated along a 250-kilometer stretch of coast) spread over about 88,000 square kilometers of land, where a minimum dose of fifteen and a maximum dose of 15,000 particles per minute were inhaled. In an experiment with bacterial spores, 600 liters of suspension were sprayed from the deck of a ship running on a three-kilometer course about three kilometers off shore, at right angles to an onshore wind. The meteorological situation was such that there was a slight tendency for the aerosol to rise and become diluted. Nevertheless, the cloud could be followed about thirty-seven kilometers, providing "infectious doses," even inside buildings, over some 250 square kilometers. Of course, the implications of a test of this sort can be downgraded in so far as it concerned a very hardy spore, but many years ago guinea pigs were successfully infected with bacteria in a more vulnerable (vegetative) form that had traveled nearly twenty-five kilometers in an aerosol. Indeed, aerosolized vegetative cells of a harmless bacterium have been made nearly as stable as the spores used in the field test mentioned earlier.

Besides a "biological decay," which is attributed to irradiation (notably by ultraviolet from the sun) or to unfavorable humidity, an infectious aerosol cloud is also subject to "physical decay." There is a progressive reduction in the effective particle concentration due to dilution, settling out under gravity, washing out by rainfall, and impact upon surfaces.

If an aerosol is generated along a line perpendicular to the wind direction, loss by lateral diffusion is significant only at the ends of the cloud; diffusion along the wind direction does not alter the total dose presented. Vertical mixing, on the other hand, may cause a rapid drop in the particle concentration. That is why an aerosol attack is most likely on a clear night when the ground loses heat by radiation and there is a "lid" of warm air to limit the upward movement of the aerosol. Very large areas could be involved by exploitation of the so-called polar outbreaks, wherein a layer of cold air a kilometer or more in thickness could carry an aerosol for hundreds of kilometers at speeds of thirty to forty kilometers per hour.

The effects of microbiological weapons, compared with their weight, are such that an attacker might easily increase the quantity released, to compensate for the decay, if he provides only "standard" protection for the active material and gives due consideration to its state at the time of release. If *F. tularensis* (rabbit fever) were aerosolized, it could perhaps be assumed that its capacity for infecting humans falls off at the same rate as it does on guinea pigs. Initially, the infectious dose for those animals is around ten to twenty cells, but it increases to about 150 to 200 when the aerosol is five and a half hours old. A relatively small amount of material would provide this dose over hundreds of square kilometers. Theoretical calculations clearly indicate the possibility of large-scale coverage, even allowing for a decay of, say, two per cent of the particles every minute. A midnight dispersion of five liters per kilometer of a suspension holding ten million million (10^{13}) particles per liter at an altitude of 100 meters along a fifty-kilometer line would—given a reasonable generator efficiency, certain meteorological conditions, and a wind speed of twenty kilometers per hour—set up a cylindrical cloud that would pass a downwind point in less than a minute. A person breathing at a rate of ten liters per minute would be exposed to about 150,000 particles. If this happened to him at 2 A.M., only 150 of those particles would still be active; in other words, the agent should have an infectious dose of not more than 150 viable particles in order to cause disease forty kilometers downwind. An individual exposed to the same cloud 120 kilometers downwind, at 6 A.M., would only contract the disease if the agent used had an infectious dose of 1.5 particles. By this time the coverage would be 6000 square kilometers.

The size of aerosol generators that would be used in limited attacks (on parliament buildings, military staff headquarters, and so on) would be so small that they could easily be concealed by a saboteur. He would certainly be vaccinated and could arrange to leave the scene in ample time before cases would start to appear; in the case of an attack with *F. tularensis*, for instance, some two to five days would elapse before the disease symptoms (fever, headache, malaise, sore throat, muscular ache, and chest pains) would begin to make themselves felt.

The conventional image of biological warfare, the covert "man with the suitcase" poisoning water supplies and ventilation systems, seems to have been discarded by many experts in the field, but this attitude may well prove to be premature, at least if one considers specific situations, for instance a sanitary breakdown due to a nuclear attack or mobilization, when the psychological repercussions of a covert biological attack might be very severe. A number of tendencies in a modern society pave the way for such attacks. Extensive and rapid communications increase the "coverage" by a lone man. Urbanization concentrates his targets in small areas. Increasing sizes in slaughterhouses, dairies, and food processing factories, and the extensive use of uniform cultivated crops, large herds of animals, and centralized fodder manufacture, make food supplies very vulnerable; similarly the development of large reservoirs increases the effectiveness of individual attacks on water supplies. Central ventilation systems in command centers, subways, cinemas, theaters, restaurants, and so on, provide a ready-made means of distributing biological agents, while society's dependence on key personnel—radar and missile operators, crews of ships, and workers in power, communications, and transport services—makes them desirable targets. Such individuals are frequently confined to environments where conventional sabotage acts might be more difficult to carry out than the introduction of a biological agent with an incubation time giving a saboteur many days to escape.

Indeed, the most disturbing aspect of biological warfare is the possibility that it might give to small groups of individuals to upset the strategic balance. It is, for instance, hard to dismiss as unrealistic an example given by Dr. Brock Chisholm, formerly head of the World Health Organization. He has speculated about a hypothetical nation making an attack on the United States by 100 vaccinated agents using botulinus toxin as the weapon. Each would import a few pounds in a body belt and proceed to one of the major cities, power sites, or military centers. At a pre-arranged time each would take a small private plane from the local airport and then dust his target from the windward side with the aid of a small, easily made apparatus. Fatalities after such an attack might range from forty to nearly one hundred per cent, and the attack might well be blamed on the U.S.S.R. Nuclear weapons would then be fired, and retaliation from the U.S.S.R. would be automatic and immediate.

Why Defence Is Difficult

The microbiological agents that might be used for offensive purposes represent a whole range of weapons systems rather than a single type of weapon. Antipersonnel, antianimal and anticrop agents all pose different defence problems; an open aerosol attack would represent one weapons system and a covert dissemination of infected insects another, quite different system. The situation is further complicated by the various means an enemy might employ in order to enhance his attack. Radiation from nuclear fallout would aggravate the results of an aerosol attack on man or animals by lowering natural resistance. Carriers like crystal needles might help viruses to penetrate plants. Elimination of chlorination would permit or simplify an attack on water supplies. And so on.

Specific prophylaxis, by vaccination for example, or cultivation of disease-resistant crop plants, has a very severe

limitation because of the great range of weapons potentially available to the enemy, who might well use microorganisms made resistant to standard prophylactic and thera-
peutic measures. Consequently, nonspecific protection by
physical and chemical means is essential. The possibilities
include: particle filtration in gas masks; sterile ventilation
in shelters and command centers at pressures above that of
the atmosphere outside; sterilization and hermetic packag-
ing in the food and beverage industries; conventional chlorination in waterworks, and so on. However, continuous
physical protection would hardly be acceptable—particularly in "peacetime," when a surprise attack might well be
launched—so an effective defence would presuppose a detection network including a sophisticated array of early-
warning, sampling, and identification devices.

One of the technical difficulties in aerosol warning is the
normal "background" of biological material in the air, and
this changes with the time and place, making it necessary
to base a setting of the alarm threshold on a thorough
knowledge of the local situation. The concentration of viable aerobic bacteria is usually less than one per liter of
outdoor air, and the protein content is around three bil-
lionths of a gram per liter. Automatic warning devices
would have to measure such minute quantities of protein
(or nucleic acid) continuously, and that involves a range of
very difficult technical problems. Also the particle content
in the narrow size range critical for infectivity would have
to be monitored.

At the present time it seems necessary to combine several principles of detection and to rely on human integra-
tion of many factors, including weather conditions, in order to issue an alarm. Many sophisticated warning systems
are conceivable, but considering the multitude of routes
available for delivery and the great number of agents that

the enemy might choose, it is highly improbable that any system will ever be able to prevent the occurrence of very significant numbers of clinical cases in the event of an attack.

Speed in the laboratory identification process is essential, since large groups of individuals would have to be detained and also because it may be too late to institute treatment by the time a clinical diagnosis is made. In the case of inhalation anthrax, treatment should start when the symptoms are vague and before the condition becomes alarming. Pneumonic plague has a mortality close to one hundred per cent if treatment is not instituted within twenty to twenty-four hours of exposure, and an early treatment is also important in the case of rabbit fever. Obviously the speed of identification required is a challenge to the microbiologist; only under exceptional conditions, where defective munitions, expended spray devices, or vector containers are recovered, would he have more than a minute amount of material available for study immediately after an attack. Microbiological laboratories themselves would be choice targets, and of course there is plenty of scope for using a multiplicity of agents to confuse the identifications.

However, biological warfare is a challenge to the microbiological profession, not only from a technical but also from an ethical point of view.

"Biological Warfare in Reverse"

A systematic study of the spread of infections in a population and surveys of certain diseases among animals provide a basis for planning defence against biological attack. A national surveillance program, involving routine immunological testing and regular surveys of insect populations, also makes possible retrospective investigations that might provide important proof to support a case of alleged biological attack—which is not an easy matter. A surveillance program might also reduce the risk of an attack being launched, particularly if aided by automatic devices. A rewarding aspect of a biological defence program is that it will improve the possibilities of coping with naturally occurring diseases. Where public health and microbiological facilities are already good, and budgets are not too tight, the problems of preferential defence (protecting some targets but not others) may not be insoluble.

In the developing regions of the world and in some smaller countries, on the other hand, the outlook is worse, and multilateral or international collaboration might be required to provide even token defence systems. In such areas the risks are probably also greater because the building of a nuclear capability is apt to be a slow process and microbiological weapons might seem to offer a tempting strategic alternative. To be of real significance defensive surveillance programs should be broad and nationwide, and they would require data handling, computer facilities, and automated analyzers beyond the reach of small or developing countries. Strangely, these countries have not so far entrusted the United Nations specialized agencies with special powers and responsibilities in respect to biological warfare.

The public health services of small countries might help to limit the spread of infections that normally respond to quarantine measures, but they could hardly cope with an aerosol cloud or a massive release of infected insects. The effects of such attacks might reach far beyond national boundaries, and effective limitation might require efforts of which many nations would be incapable. Such countries should be able to buy an "insurance policy" that would, for instance, guarantee expert advice and immediate deliveries by air of substantial quantities of vaccines, antibiotics, and chemicals in the event of an attack.

Several of the United Nations specialized agencies have a certain competence in the biological defence area. The World Health Organization (W.H.O.), for instance, can be expected to have an interest in the public health aspects. The Food and Agriculture Organization (F.A.O.) should be concerned with the protection of plants and animals, while the relevant microbiological research, documentation, and science policy falls within the scope of UNESCO. In view of the political dangers of impromptu international committees appointed to study allegations of biological attack, the establishment of an independent International Microbiological Agency as a parallel to the International Atomic Energy Agency (I.A.E.A.) has recently been proposed. This would tie a control function in biological warfare to the peaceful applications of microbiology, in the same way as the control of reactor fuels is tied to the peaceful uses of nuclear energy within the I.A.E.A. Since biological warfare could well upset the delicate power balance that now affords a precarious stability in the world, the superpowers should have a very real interest in biological disarmament and control. They are in the best position to know that the biological weapons are likely to remain erratic and difficult to handle as part of military games theory, the more so the cruder the technology. The superpowers should also know that a comprehensive defence program would involve almost prohibitive costs; they should also be conscious of the drain of such a program on a professional group (the microbiologists) that has many vital functions in society and is of considerable significance for the technical aid to developing countries that is now an important part of their foreign policy.

Microbiological weapons come from some of the largest

and best operated establishments for applied microbiology in the world of today, and unlike other military establishments they would require no major reorganization to switch most of their activities to a highly constructive path. Public health could be "biological warfare in reverse," because most of the efforts that go into the development of weapons could equally well be applied to the development of new vaccines, new administration techniques, and other methods relevant to prevention of disease, which is such an important factor in the industrialization process of the developing countries. It is true that "biological warfare in reverse" would tend to accelerate the population explosion, but it might also provide new answers to some problems in family planning (for example, through immunological control of sperm production). To deny its potential for good would be to deny the impact of antibiotics on public health and husbandry, of biological insect control and seed inoculation on agriculture, and of the fermentation technology and food microbiology on the production of nutrients.

Realizing that applied microbiology is of great significance to the developing countries of the world, UNESCO has launched an ambitious microbiological program, and a number of countries have included microbiological efforts in their aid programs. If their significance can be conclusively demonstrated, and if the desire to benefit becomes strong enough in the developing countries, the latter might be induced to accept biological-weapons control and inspection in return for assistance, by analogy with provision of fissile materials through the I.A.E.A.

Different types of treaties for control of biological weapons have been proposed but, as for other weapons, there is an unwillingness to accept treaties if they do not embody the principle of adequate inspection. There is a consensus of opinion expressed in the Soviet and Western proposals for general and complete disarmament, to the effect that biological and chemical weapons should eventually be subject to control. Insurance against the use of biological weapons in a disarmed world might, however, prove to be impossible owing to the many delivery systems available to the potential user. Inspections to check on the possible manufacture of biological weapons also involve severe problems. Techniques include registration of scientific personnel, accounting of materials, fiscal controls, and visits to conceivable production and testing facilities. If an efficient system for the regulation of research and development, and for preventing testing and clandestine stockpiling cannot be proposed, it is unlikely that nations will be prepared to disarm. The Pugwash movement early recognized this problem and in recent years has operated a special study group on biological warfare, which has organized a series of experimental inspections in four countries, in order to gain practical control experience. This study has not yet been completed, but the inspectors have come to the conclusion that the possibilities of arriving at a reasonably effective control system are better than originally expected.

Control and inspection are hardly possible with regard to "sabotage" quantities of biological weapons, but large-scale military efforts are probably not easy to hide. For instance, the Pine Bluff Arsenal in Arkansas, where biological weapons are produced, as well as toxic-chemical and riot-control munitions, covers some 15,000 acres. Facilities for field testing would seem to be even more difficult to conceal. The Dugway Proving Ground is an example, occupying an area in Utah larger than the state of Rhode Island. Strong sunlight might help to accelerate the biological decay in a testing ground, but animals might harbor the microorganisms, and some bacteria might be so resistant that they would be detectable by normal sampling techniques over long periods and at considerable distances from the test site. A high degree of resistance has, for instance, been observed in the case of anthrax, which was used for biological warfare experiments on the small island of Gruinard, off the northwest coast of Scotland, during World War II. After a recent study, it was stated that this island "may remain infected for one hundred years."

Nevertheless, control and inspection problems remain daunting because there is nothing about a production facility that would necessarily show on the outside of the buildings, and extensive field tests might be staged covertly on foreign soil. Consequently an intelligence activity would have to be part of the effort. For that to be fully effective, unorthodox supporting initiatives might also be required —perhaps internationalization of the microbiological profession, or provisions for diplomatic immunity and awards for disclosure of military preparations contrary to international agreement.

Fortunately, from the political point of view it might not be necessary to have an absolutely "leak-proof" system. The most important aspects would be found in side effects. The preparations for a control and inspection system would require contacts that would help to reduce the mistrust among nations, and they would tend to reinforce the international conscience in the field. International and regional agreements aimed at neutralizing the political dangers of allegations of biological attack and at limiting the spread of weapons technology would then logically follow.

Moral and Military Thin Ice

Discussions of the relative moral merits of napalm burning, saturation bombing, and nuclear warfare versus the use of infectious aerosols seem pointless without a careful balancing of the long-range interests in terms of self-expression and happiness as far as both the attacker and the target population are concerned. Such a type of penetrating analysis is difficult for anybody to perform, particularly for the military man or the average politician. Superficially the microbiological weapons might seem preferable, because they include a wide range of agents that temporarily incapacitate rather than kill human beings and they do not have the obvious genetic effects that we associate with nuclear weapons. However, this is hardly enough to open the door to their use, since the young, the elderly, and the infirm may be killed by so-called incapacitating agents and there is in any case no clear dividing line between those weapons and other agents giving a very high mortality. Once the use of microbiological weapons became acceptable, the supposed "humane" aspects would be offset by an almost inevitable escalation.

Future conflicts are likely, in short, to breed weapons modeled upon nature's own ecological system. Man's success with biological control of insects and other pests indicates a road that is certainly appalling. Our dismay should not be aroused merely because microbiological agents are invisible and lack smell and taste, or because a perfect defence will never exist except in the minds of some theorists. What should determine our attitude is rather the fact that these weapons, like other weapons of mass destruction, will never be selective enough to spare individuals who are not responsible for the situation that breeds them. Meanwhile, the war gamesmen who balance the costs of antiballistic missiles (ABM) against "acceptable" megadeaths seem to walk on thin ice in presuming fair play or Expected Gentlemanly Behavior (EGB) by the opponent, as far as biological weapons are concerned.

AUTHOR'S NOTE: For a more comprehensive discussion, in particular of defence problems, the reader is referred to an article by the author in *Annual Review of Microbiology* (Palo Alto, California: Annual Reviews, Inc., 1967); there the references to the relevant literature will also be found.

ROBOTS ON THE MARCH

B Y

M. W. THRING
United Kingdom

Professor Thring is head of the mechanical engineering department of Queen Mary College, University of London. He is an authority on high-temperature engineering, but he has also spent several years developing practical robots for peaceful purposes.

ROBOTS that conform with the long-standing image of science fiction, that is to say free-moving, self-regulating, active vehicles, are now in a fairly advanced stage of development. Their immediate application is in carrying out limited but complex tasks in the office, home, and factory. They can, however, be adapted to military purposes, to operate fighting machines without human crews.

The characteristic elements of a robot are:

1. a computer, the functions of which are first, to remember the detailed instructions of the duties the machine is to perform and, second, to adapt these duties in prede-

termined ways according to its observations of external situations;

2. senses, which may include response to touch, sight (by light, radar, infrared), and sonar, and from which the computer can recognize a limited number of categories of objects;

3. arms and hands to operate other machines and manipulate objects;

4. means whereby the robot propels itself about the region in which it is to work.

The most difficult of these elements to develop is the ability of the machine to recognize objects by sight, sonar, or touch. But within a few years this problem will be completely solved. With the concurrent progress in miniaturization of electronics, a robot smaller than a man will be available, capable of any small range of selected specialized tasks that a man can do. Being a robot, however, it will have no feelings of fear or any desire to avoid its own destruction, except such reflexes as the designer may incorporate to preserve his machine. The robot can be much faster and more accurate than a man in taking any predetermined action when the external situation provokes it, in accordance with the machine's instructions.

Once the computer and sensing devices have been developed, the complete robot control system—suitable for a ground weapon carrier (robot infantryman), a tank, a ship or submarine, an airplane, or rocket—could be mass-produced at a cost of, say, $10,000. Although this may be greater than the value set by a country on its human soldiers, one-for-one comparisons are not really appropriate. The advantages of the increased speed and accuracy of the robot, of the increased effectiveness of the carrying device, and of the reduced size and vulnerability of the unmanned

vehicle, would make the human soldier completely obsolete. One robot would perform all the functions of the
crew of a bomber or tank or submarine and can be built in
as part of the unit, requiring no space for moving about.

The application of such robots to war purposes will
plainly change the whole character of war. The main effect
will come from the development of "suicide" bomb carriers of incredible accuracy and ranges of thousands of kilometers. Before we consider those, we can look at the use
of robot soldiers in more conventional war situations.

Robot Soldiers

The traditional function of a soldier, whether on foot
or in a vehicle, is to find the enemy soldier and destroy him
with a weapon. To do so he must propel himself across the
land until he is close enough to detect the enemy by sight
and aim his weapon. All these functions can be performed
far better by a compact armored robot with a sophisticated
walking mechanism. It can be equipped with light and infrared vision, radar, and sonar; it can carry a built-in weapon
such as a gun or rocket launcher, directed by computer
to an exact range and aim on an enemy; and the same computer can steer the robot, even across terrain requiring cir-
cumambulations, to the desired point, and maneuver it so
as to search for the enemy. With a laser or plasma-torch
light spot it can blind all human eyes looking at it. The armored robot can carry enough fuel to run for a week without stopping, and it can operate with comparative impunity in highly radioactive regions. Sensitive parts such as the
computer and the sensory detectors can be placed in well
armored compartments close to the ground to make them
almost invulnerable. The light and other signals can, if

appropriate, reach the shielded detectors via tall periscopes or external aerials.

A regiment of such robot soldiers could be controlled by a human officer in a tank who remains several kilometers behind them in a well-protected spot. A line of such robots spaced twenty meters apart might be deployed to move at fifteen kilometers per hour through a jungle and destroy all men encountered there. A bigger robot vehicle could carry a horizontal chain saw in front of it and cut down all trees and push them to either side so as to clear a straight track ten meters wide through the jungle, which could then be patrolled by robot soldiers to ensure that no man would be able to cross the track. The robot soldiers might be re-fueled and rearmed by service robots traveling along such a track once a week.

In conditions where antipersonnel weapons and radio-active fallout make battlefields untenable by human troops, robots may be the only means of carrying out the tradi-tional military task of occupying and holding ground. They will also be immune to chemical and biological poisons, to napalm, and most other weapons intended to kill human-ity. One can envisage battles, or whole campaigns, waged between opposing robot armies, with comparatively little human participation.

The task of policing a terrorist-ridden town might be left almost entirely to specially designed robots that would pa-trol the streets, send television pictures of any event to a central control station, and fire an explosive shell at any gunflash with a delay of only a fraction of a second.

Robots Instead of Heroes

Manned fighting vehicles have reached a very ad-vanced stage of development: they include the tank on land, the submarine and aircraft carrier on the sea, and bombers and fighters in the air. As long as men are present in a fighting vehicle, it is a matter of prime necessity that, after the vehicle has done its fighting, it should be able, in principle at least, to bring the men safely back to base. This requirement severely restricts the destructive payload carried by the vehicle and halves the distance from its base at which it can deliver its payload. Great attention has to be paid to safety factors in design, to human comfort, and to durability.

If a machine can be made to fight without any crew, then it can be designed solely for its destructive task—typically to deliver the maximum possible explosive charge to the appropriate target in the enemy territory. There is no need for the machine to return, provided only that it should do damage of considerably greater value than that of the machine itself. In this case the duties of the controlling computer are primarily those of navigation and steering the machine to the target over a distance of some hundreds or thousands of kilometers from its starting point, and of arriving within a few meters of the required spot before it detonates its explosive charge—perhaps a nuclear bomb. The machine has to take avoiding action, if necessary, from weapons attempting to stop it, but it has to find its way back on its path in spite of such evasion. It may also have to overcome obstacles in the form of barriers placed in its path. It can do more accurately and more rapidly than humans any duty that the crews perform in present machines, and it requires no courage to carry out the task it was designed and built to do.

We can already make nuclear explosives to destroy every human being and every important building and piece of equipment on the entire land surface of the earth. The chief problem is to deliver them accurately to towns, large buildings, and all groupings of humans or equipment. Manned bombers and ballistic missiles are rather unreliable for this purpose, unless the bomb is sufficiently large to allow for a considerable margin of error. Moreover, now that rapidly responding interception systems are becoming available operationally against aircraft and in prospect against ballistic missiles, speed is no longer the chief guarantee of immunity from interception. Slower but virtually indestructible machines that can take their time in finding their targets, and do so with precision, may well have the advantage. Those machines operating on the enemy's own ground, or at low altitudes above it, will benefit from his inhibitions against using nuclear weapons to stop them in such circumstances. And the psychological effect of a swarm of unstoppable robots bearing ponderously down upon their targets might well force capitulation, even before any of them had exploded.

The technical requirements can be met completely by the development of robot pilots for bomb-delivery machines, on land, in the air, in outer space, and under water.

The Walking Bombs

For our central concept, let us take a one-hundred-ton tank, shaped like a tortoise with outer steel shells above and beneath it, and traveling on legs. The walking device is the "centipede" in which the vehicle is supported on five to ten rubber-shod legs on each side. An engine drives two stout chain loops running the full length along each side of the vehicle, the chains being mounted one above the other within the protective shell. Both chains on one side are driven at the same speed, which is the forward speed of the vehicle. The upper end of each leg is attached to a link in the upper chain, and the middle of the leg is attached to the link of the lower chain, which is always vertically below the corresponding point on the upper chain. Thus the legs are always held perpendicular to the chains; they protrude through a slot in the lower shell. The legs support the vehicle while it runs forward on the chains; then the rear of the vehicle catches up with a leg, the leg rises off the ground and moves forward to the front. There it travels downward by a distance equal to the diameter of the front sprocket wheels and so can, if need be, hoist the body up a vertical step of any height less than this diameter, which might be 1.5 meters. The leg-chains are driven via clutches and a three-speed gearbox, for "running" (60 kilometers per hour), "walking" (8km/h), and obstacle climbing (very slow).

The lower shell is beveled away at the front so that the full downward movement of the legs can be used to lift the whole "tank" onto an obstacle of this height. The legs are thus exposed by this amount at the front and by about 0.5 meter as they rest on the ground for the main part of the movement. However, these legs are solid high-tensile steel bars, perhaps fifteen centimeters thick, so that they are extremely difficult to damage by gunfire or mines; all other moving parts are of similar ruggedness. Anything else that has to protrude from the shell, for example an aerial for radar observation, will be made retractable. For clearing obstacles up to ten meters high, the machine may carry four telescopic legs operated as hydraulic jacks.

It can also have a small gun firing forward for destroying obstacles in its path, and it can have a flail device for blowing up a mine field in front of it. The "tank" is, however, so heavily armored that it can go over ordinary explosive mines with impunity. Even if it is thrown into the air it will roll the right way up and continue its path. It can also crawl any distance under six meters of water by extending its air-intake tube above the surface, or travel a considerable distance under much deeper water by using a bottle of compressed air or oxygen. The latter provision would also give the machine time to dig itself out if it were buried by an explosion.

The body of the machine is mostly solid steel, with only small spaces deep inside it for the engine, fuel, and control system. An internal-combustion engine developing 200 horse-power, with enough fuel to propel the machine up to 3000 kilometers, would be appropriate. (Alternatively, it is possible to arrange for the nuclear fission component of the bomb itself to work on the journey in a condition of controlled release of heat, heating air or boiling water to drive a turbine for propulsion.)

The computer and the engine occupy separate hollow spaces inside the main steel body, each approximately 0.5 meter in diameter and one meter long. The computer has to navigate by dead reckoning based on gyrocompass readings and exact measurements of distance covered, and checking its position by radar observations of the terrain and man-made structures. If need be, navigation by sun and stars can be added as a third system. Should it encounter an obstacle taller than it can climb over, the machine must cast round it to find a way past and then automatically revert to the track that it has left. As far as possible, however, it will be set on a course that avoids large towns and cliffs on its way to its objective.

The computer is programed with maps of the route it is to follow, and given the detailed information of how to recognize the landmarks along the way by radar. It is fully instructed about when to use different speeds, how to overcome or avoid various obstacles or traps, and how to detect and traverse mine fields. It can achieve as accurate a placing of its bomb as could a man driving a tank into Times Square or Red Square, because the final location of the target will be by radar or optical recognition, based on comparison with pictures stored in the memory.

On the outbreak of war a fleet of one hundred or more such walking tanks, loaded with their nuclear bombs, is dispatched across a frontier or landed from manned transport submarines at a series of suitable points along a coast, up to 3000 kilometers from each of the cities or other major assets of the enemy. They set off, responding to circumstances with the speed and accuracy of the computer and without any emotional interference from human feelings of uncertainty, danger, or suicide. Each "tank" that reaches its target carries sufficient nuclear explosive to destroy it entirely, but without unnecessary "overkill." In the not unlikely event of war being stopped by the capitulation of the enemy before they reach their targets, the machines could be immobilized and their bombs set to safety by a suitable combination of radio signals. That is the only circumstance in which communications from the rear would be needed, and plainly it would then be in the enemy's own interest to refrain from jamming.

What can the enemy do to stop the "tanks"? Conventional antitank ditches and parapets would have to be very wide and deep (more than ten meters) to trap them, and therefore extraordinarily expensive to provide around a large perimeter; they could not be built overnight. Even if they existed, it would be a simple matter to expend a few of the tanks in blasting passages for the remainder. Armor-piercing shells would be almost useless against the solid hull. One of the two most vulnerable parts of the system is the chain drive for the legs, especially at the front sprockets, but that would be built so massively that it would survive even if the enemy contrived to shoot powerful shells upwards through the leg slots—perhaps using a modified land mine, or by getting underneath when the telescopic legs were extended.

The other vulnerable feature is the sensory system. The radar aerial would be exposed for only a brief scan at intervals, but a lucky shot could carry it away. Reserve aerials would minimize the risk of "blinding" in this fashion, although all would be susceptible to jamming. It is, however, important to realize that the radar is only an aid to navigation and not indispensable. The navigation by dead reckoning will be very accurate, and even a blinded "tank" can cope in a logical manner with obstacles, which it detects simply by being stopped by them. By the same token, the computer will recognize any attempt to confuse it by false radar echoes that alter its landscape; its first reliance will be on its dead reckoning and it will "know" what the landscape should look like.

The only reliable way of stopping the "tank" would be with a direct or very near hit with a nuclear weapon. To the natural inhibitions against using even small nuclear weapons for defence on one's own territory is added a much greater hazard if the machine is carrying a large H-bomb. The latter can easily be programed to explode when a restraining signal from the computer ceases, signifying that the machine is "dead." It is, of course, preferable to have one's rural areas rather than one's cities devastated in this way, but by suitable choice of routes through demographically and economically important areas the attacker can probably make deliberate explosion of the machines unacceptable as a means of defence. Therein lies the attacker's great advantage; however the defender tries to cope with the machines—even, perhaps, by building comparable robots to tussle with the robot bomb-carriers—the very best he can expect, for all his pains, is a thermonuclear explosion away from the target but still on his own territory. My belief is that everything else would be tried, in a frenzy of improvisation, and that many of the "tanks" would reach their targets.

The very slowness of the "tanks," compared with attack by rocket or bomber, alters a common assumption about nuclear weapons—that there is a bi-stable situation in which the bombs are either held back or delivered more or less instantly, with virtually no time for second thoughts on either side before the bombs go off. We can now see an intermediate state of affairs, in which the bombs are plainly and relentlessly on their way, constituting a threat of the gravest kind but still allowing perhaps several hours for the political decisions that can halt them before the first reaches its target.

Other Robot Weapons Systems

Compared with the walking "tanks," other robot weapons systems will have over-all characteristics less unfamiliar to us. They will generally enhance the performance of systems at present falling within the scope of guided weapons on the one hand and manned aircraft and submarines on the other, by combining the advantages of both: giving the guided weapon something of the reasoning power of men or, conversely, giving the systems at present operated by men the fast response, compactness, and expendability that characterize the guided missile.

Present ballistic missiles are crude compared with the robot-controlled nuclear warhead of the future. It will be steered about its general ballistic path, using ancillary rockets, by a miniaturized computer responsible for navigation, evasion of antiballistic missiles, and eventual target identification. It will constantly reckon the fuel and time remaining for it to recover its required path, and will make such random alteration of course as the circumstances allow, to confuse the defences.

Again, imagine a robot-controlled supersonic aircraft designed to carry a megaton bomb for 5000 kilometers. It will take avoiding action whenever interception is attempted, by very rapid random turns, climbs, and dives, and by sudden increases of speed at random intervals, using boost rockets or ramjets. The computer operates the flight controls. The machine flies at great height until it approaches the region where, as indicated in its program, it must expect enemy defences to be active; then it comes down very low and flies thirty to 150 meters above the ground, navigating with the electronic sensors and other aids, including maps, that a human crew would use, but responding to incoming information very much faster. Accelerations in maneuver can be much more violent than humans could withstand. When the aircraft detects the target it has been instructed to attack, it explodes its nuclear bomb (and itself) at the exact height and position over the target where it will do maximum damage.

Simpler robot-controlled "suicide" bombers can carry conventional explosives. They will be adequate for dealing with compact targets such as ships and will, for example, seek out and crash on the deck of any ship operating in a specified area of the ocean that contains no vessels of the attacking side. Other robot-controlled aircraft will fly at subsonic speed one hundred meters above the ocean trailing antisubmarine detectors on paravanes a few meters below the water; as soon as such an aircraft detects a submarine by sonar it will locate its depth and position by circling round and making magnetic measurements. Nuclear depth charges set and dropped by the robot aircraft will then destroy the submarine.

On the other hand, robot submarines may be difficult to detect and destroy. Take the men out of a submarine and there ceases to be any problem of providing space, comfort, warmth, or good air; it becomes easy to design for operation at great depths in the sea, as the pressure inside it can be equal to the outside sea pressure. Even the robot controller and the nuclear power source can be made to work perfectly well at high pressures. The craft can also withstand much greater punishment under attack than a manned submarine.

A small robot submarine, carrying an H-bomb and capable of finding its way for a distance of 3000 kilometers at depths of up to 600 meters or more, can perform a function similar to that of the landborne "tank" discussed earlier. Navigating partly by dead reckoning and partly by sonar, flotillas of such submersibles would locate enemy ports and coastal installations and destroy them, if need be after forcing or blasting their way through any gates or defensive nets.

Other computer-controlled submarines could have a naval role. They would scour assigned parts of the oceans for all shipping, including hostile submarines, aircraft carriers, and cargo vessels. They would detect the ships by sonar and magnetic sensors, and sink them by ramming at high speed, coming up from a great depth as they did so. Unlike Jules Verne's Nautilus, which used a similar tactic, they would have no humanity-hating captain aboard but only a robot controller programed to destroy. They would have a normal maximum speed at depth of fifty km/h but a crash speed for ramming of 150 km/h; their size would be very small, set by the size of the smallest nuclear reactor, the hull of the submarine being the only shielding necessary. Such robot hunter submarines would be programed to sink every surface ship or other submarine in fixed areas excluding narrow lanes for use by friendly ships.

War of the Robots

Present-day automatic guidance and control systems for weapons are still comparatively primitive, and lack the versatility and capacity for tactical judgment of human servicemen. It is therefore still possible to argue the importance of manned aircraft, of infantry soldiers and traditional warships. Within a very few years this situation will be completely altered. To the already indisputably faster response of electronic control will be added as much capacity for information storage, decision-making, sensory input, and pattern-recognition as operational staffs care to specify. Men will cease to be valued in battle; on the contrary, they will be recognized as a grave complication in systems design, introducing great penalties of volume, weight, and vulnerability. Indeed, once the assumption that war is an affair for humans has been shaken, the military incentives to develop robot weapons will become irresistible. Robot soldiers immune to flying metal and to nuclear flash and radiation are simply a better military proposition than human troops, however brave and resourceful. Aircraft that need not get home, submarines that require no pressure-proof hull, and yet which can carry out all the maneuvers and opportunistic functions of the corresponding manned fighting vehicles, can have a performance in relation to cost that the latter could never match.

As the chances of human survival in battle dwindle toward zero, with the deployment of weapons that leave little to chance, humans are likely, in future wars, to stand helplessly by as a struggle rages between robot armies, navies, and air and rocket forces. To suppose that humanity will benefit by leaving the fighting to the machines is, however, to ignore the fact that many of the robot weapons will be carrying weapons of mass destruction targeted on human populations.

HOW TO WRECK THE ENVIRONMENT

B Y

GORDON J. F. MacDONALD
United States

Professor MacDonald is associate director of the Institute of Geophysics and Planetary Physics at the University of California, Los Angeles. His researches have embraced a remarkable diversity of natural phenomena, and his professional interests are further extended by his participation in national science policy-making. He is a member of President Johnson's Science Advisory Committee.

AMONG FUTURE MEANS of obtaining national objectives by force, one possibility hinges on man's ability to control and manipulate the environment of his planet. When achieved, this power over his environment will provide man with a new force capable of doing great and indiscriminate damage. Our present primitive understanding of deliberate environmental change makes it difficult to imagine a world in which geophysical warfare is practiced. Such a world might be one in which nuclear weapons were effectively banned and the weapons of mass destruction were those of environmental catastrophe. Alternatively, I can envisage a world of nuclear stability resulting from parity in such weapons, rendered unstable by the development by one nation of an advanced technology capable of modifying the earth's environment. Or geophysical weapons may be part of each nation's armory. As I will argue, these weapons are peculiarly suited for covert or secret wars.

Science-fiction literature contains many suggestions of how wars would progress if man indeed possessed the ability to change weather, climate, or ocean currents. Many of these fictional suggestions, and other more serious discussions, fail to take into account the limitations of nature. Jules Verne gave a detailed discussion of displacing the earth's polar caps, thus making the world's climatic zones more equitable (*Les Voyages Extraordinaires; Sans Dessus Dessous*, Metzel, 1889). Verne's proposal was to eliminate the twenty-three-degree tilt in the earth's axis, putting it at right angles to the sun-earth plane. However, as Verne correctly pointed out in a subsequent discussion, the earth's equatorial bulge stabilizes our planet, and even the launching of a 180,000-ton projectile would produce a displacement of only one-tenth of a micron. Senator Estes Kefauver, Vice-Presidential candidate in the 1956 American election, rediscovered Verne's original proposal and was seriously concerned with the tipping of the earth's axis. He reported that the earth's axis could, as the result of an H-bomb explosion, be displaced by ten degrees. Either Senator Kefauver or his scientific advisers neglected the stabilizing influence of the earth's bulge. The maximum displacement that can be expected from the explosion of a one-hundred-megaton H-weapon is less than one micron, as Walter Munk and I pointed out in our book, *Rotation of the Earth* (Cambridge University Press, New York, 1960).

Substantial progress within the environmental sciences is slowly overcoming the gap between fact and fiction regarding manipulations of the earth's physical environment. As these manipulations become possible, history shows that attempts may be made to use them in support of national ambitions. To consider the consequences of environmental modification in struggles among nations, we need to consider the present state of the subject and how postulated developments in the field could lead, ten to fifty years from now, to weapons systems that would use nature in new and perhaps unexpected ways.

The key to geophysical warfare is the identification of the environmental instabilities to which the addition of a small amount of energy would release vastly greater amounts of energy. Environmental instability is a situation in which nature has stored energy in some part of the earth or its surroundings far in excess of that which is usual. To trigger this instability the required energy might be introduced violently by explosions or gently by small bits of material able to induce rapid changes by acting as catalysts or nucleating agents. The mechanism for energy storage might be the accumulation of strain over hundreds of millions of years in the solid earth, or the supercooling of water vapor in the atmosphere by updrafts taking place over a few tens of minutes. Effects of releasing this energy could be worldwide, as in the case of altering climate, or regional, as in the case of locally excited earthquakes or enhanced precipitation.

Weather Modification

The earth's atmosphere is an envelope of air that rotates, for the most part, at the same speed as the underlying continents and oceans. The relative motion between the atmosphere and the earth arises from sources and sinks of energy that vary in location and strength but which have, as their ultimate source, the sun's radiation. The quantities of energy involved in weather systems exceed by a substantial margin the quantity of energy under man's direct control. For instance, the typical amount of energy expended in a single tornado funnel is equivalent to about fifty kilotons of explosives; a single thunderstorm tower exchanges about ten times this much energy during its lifetime; an Atlantic hurricane of moderate size may draw from the sea more than 1000 megatons of energy. These vast quantities of energy make it unlikely that brute-force techniques will lead to sensible weather modification. Results could be achieved, however, by working on the instabilities in the atmosphere.

We are now beginning to understand several kinds of instabilities in the atmosphere. Supercooled water droplets in cold clouds are unstable, but they remain liquid for substantial periods of time unless supplied with nuclei on which they can freeze. Conversion of water droplets to ice through the introduction of artificial nuclei can provide a local source of energy. This released heat can cause rising air currents, which in turn lead to further formation of supercooled water. This process may lead to rainfull at the ground greater than that which would have been produced without the artificial nucleation. A second instability may arise, in which water vapor condenses into water, again affecting the distribution of sensible energy. On a larger scale, there is the so-called baroclinic instability of atmospheric waves that girdle the planet. Through the imbalance of heat between equator and pole, energy in this instability is stored, to be released in the creation of large cyclonic storms in the temperate zones. There are other, less well understood instabilities capable of affecting climate; I shall return to them later.

What is the present situation with respect to weather modification and what might be reasonably expected in the future? Experiments over the past eighteen years have demonstrated unequivocally that clouds composed of supercooled water droplets can be transformed into ice-crystal clouds by seeding them with silver iodide, "dry ice" (frozen carbon dioxide), and other suitable chemical agents. This discovery has been applied operationally in the clearance of airports covered by supercooled ground fog. No analogous technique has yet evolved for clearing warm fog, although several promising leads are now being investigated. In the case of warm fog, the atmospheric instability is that water vapor distributed in small drops contains more surface energy than the same water distributed in large drops. The trick for clearance of this warm fog will be to discover some way of getting the small drops to organize themselves into larger ones and then fall to the ground.

There is increasing, though inconclusive, evidence that rainfall from some types of clouds and storm systems in temperate regions can be increased by ten to fifteen per cent by seeding. Somewhat more controversial evidence indicates that precipitation can be increased from tropical cumulus by techniques similar to those employed in temperate regions. Preliminary experiments on hurricanes have the aim of dissipating the clouds surrounding the eye of the storm in order to spread the energy of the hurricane and reduce its force. The results are controversial but indicate that seeding can, in certain circumstances, lead to a marked growth in the seeded cloud. This possibility may have merit in hurricane modification, but experimentation has not yet resulted in a definitive statement.

Regarding the suppression of lightning, there is mixed but largely promising evidence that the frequency of cloud-to-ground strokes can be reduced by the introduction of "chaff" strips of metallic foil of the kind used for creating spurious echoes in enemy radars.

In looking to the future, it is quite clear that substantial advances will be made in all of these areas of weather modification. Today, both military and civilian air transport benefit from progress in the clearance of ground fog. Further progress in the technology of introducing the seeding agent into the fog makes it likely that this type of fog dispersal will become routine. In a sense, fog clearing is the first military application of deliberate manipulation of weather, but it is, of course, very limited.

Large field programs are being undertaken in the United States to explore further the possibility of enhancing precipitation, particularly in the western and northeastern states. On the high ground of the western states, snow from winter storms provides much of the country's moisture. Investigations are under way to see if seeding can lead to an increased snowpack and thus enhance the water resources. Intense interest in this form of weather modification, coupled with an increased investigation of the physics of clouds, is likely to lead to effective cloud modification within the next five to fifteen years. At present the effects are measured only statistically, and too little has been done in cloud observation before and after seeding in the way of precisely pinpointing which clouds are most likely to be affected.

As far as military applications are concerned, I conjecture that precipitation enhancement would have a limited value in classical tactical situations, and then only in the future when controls are more thoroughly understood. One could, for example, imagine field commanders calling for local enhancement of precipitation to cover or impede various ground operations. An alternative use of cloud seeding might be applied strategically. We are presently uncertain about the effect of seeding on precipitation down wind from the seeded clouds. Preliminary analysis suggests that there is no effect 200-300 miles down wind, but that continued seeding over a long stretch of dry land clearly could remove sufficient moisture to prevent rain 1000 miles down wind. This extended effect leads to the possibility of covertly removing moisture from the atmosphere so that a nation dependent on water vapor crossing a competitor country could be subjected to years of drought. The operation could be concealed by the statistical irregularity of the atmosphere. A nation possessing superior technology in environmental manipulation could damage an adversary without revealing its intent.

Modification of storms, too, could have major strategic implications. As I have mentioned, preliminary experiments have been carried out on the seeding of hurricanes. The dynamics of hurricanes and the mechanism by which energy is transferred from the ocean into the atmosphere supporting the hurricane are poorly understood. Yet various schemes for both dissipation and steering can be imagined. Although hurricanes originate in tropical regions, they can travel into temperate latitudes, as the residents of New England know only too well. A controlled hurricane could be used as a weapon to terrorize opponents over substantial parts of the populated world.

It is generally supposed that a hurricane draws most of its energy from the sea over which it passes. The necessary process of heat transfer depends on wave action that permits the air to come in contact with a volume of water. This interaction between the air and water also stirs the upper layers of the atmosphere and permits the hurricane to draw on a substantially larger reservoir of heat than just the warm surface water. There may be ways, using mono-molecular films of materials like those developed for covering reservoirs to reduce evaporation, for decreasing the local interaction between sea and air and thus preventing the ocean from providing energy to the hurricane in an accelerated fashion. Such a procedure, coupled with selective seeding, might provide hurricane guidance mechanisms. At present we are a long way from having the basic data and understanding necessary to carry out such experiments; nevertheless, the long-term possibility of developing and applying such techniques under the cover of nature's irregularities presents a disquieting prospect.

Climate Modification

In considering whether or not climate modification is possible, it is useful to examine climate variations under natural conditions. Firm geological evidence exists of a long sequence of Ice Ages, in the relatively recent past, which shows that the world's climate has been in a state of slow evolution. There is also good geological, archaeological, and historical evidence for a pattern of smaller, more rapid fluctuations superimposed on the slow evolutionary change. For example, in Europe the climate of the early period following the last Ice Age was continental, with hot summers and cold winters. In the sixth millennium B.C., there was a change to a warm humid climate with a mean temperature of five degrees Fahrenheit higher than at present and a heavy rainfall that caused considerable growth of peat. This period, known as a climatic optimum, was accentuated in Scandinavia by a land subsidence that permitted a greater influx of warm Atlantic water into the large Baltic Sea.

The climatic optimum was peculiar. While on the whole there was a very gradual decrease of rainfall, the decrease was interrupted by long droughts during which the surface peat dried. This fluctuation occurred several times, the main dry periods being from 2000 to 1900, 1200 to 1000, and 700 to 500 B.C. The last, a dry heat wave lasting approximately 200 years, was the best developed. The drought, though not sufficiently intense to interrupt the steady development of forests, did cause extensive migrations of peoples from drier to wetter regions.

A change to colder and wetter conditions occurred in Europe about 500 B.C. and was by far the greatest and most abrupt alteration in climate since the end of the last Ice Age. It had a catastrophic effect on the early civilization of Europe: large areas of forest were killed by the rapid growth of peat, and the levels of the Alpine lakes rose suddenly, flooding many of the lake settlements. This climatic change did not last long; by the beginning of the Christian era, conditions did not differ greatly from current ones. Since then climatic variations have continued to occur, and although none has been as dramatic as that of 500 B.C., a perturbation known as the little ice age of the seventeenth century is a recent noteworthy example. The cause of these historical changes in climate remains shrouded in mystery. The rapid changes of climate in the past suggest to many that there exist instabilities affecting the balance of solar radiation.

Indeed, climate is primarily determined by the balance between the incoming short wave from the sun (principally light) and the loss of outgoing long-wave radiation (principally heat).

Three factors dominate the balance: the energy of the sun, the surface character of terrestrial regions (water, ice, vegetation, desert, etc.), and the transparency of the earth's atmosphere to different forms of radiated energy. In the last connection, the effect of clouds in making cool days and relatively warm nights is a matter of familiar experience. But clouds are a manifestation rather than an original determinant of weather and climate; of more fundamental significance is the effect of gases in the atmosphere, which absorb much of the radiation in transit from the sun to the earth or from the earth into space. Intense X-rays and ultraviolet from the sun, together with high-energy atomic particles, are arrested in the upper atmosphere. Only the narrow band of visible light and some short radio waves traverse the atmosphere without serious interruption.

There has been much controversy in recent years about conjectured over-all effects on the world's climate of emissions of carbon dioxide to the atmosphere from furnaces and engines burning fossil fuels, and some about possible influences of the exhaust from large rockets on the transparency of the upper atmosphere. Carbon dioxide placed in the atmosphere since the start of the industrial revolution has produced an increase in the average temperature of the lower atmosphere of a few tenths of a degree Fahrenheit. The water vapor that may be introduced into the stratosphere by the supersonic transport may also result in a similar temperature rise. In principle it would be feasible to introduce material into the upper atmosphere that would absorb either incoming light (thereby cooling the surface) or outgoing heat (thereby warming the surface). In practice, in the rarefied and windswept upper atmosphere, the material would disperse rather quickly, so that military use of such a technique would probably rely upon global rather than local effects. Moreover, molecular material will tend to decompose, and even elemental materials will eventually be lost by diffusion into space or precipitation to the surface. At intermediate levels, in the stratosphere, materials may tend to accumulate, though the mixing time for this part of the atmosphere is certainly less than ten years and may be a few months. If a nation's meteorologists calculated that a general warming or cooling of the earth was in their national interest, improving their climate while worsening others, the temptation to release materials from high-altitude rockets might exist. At present we know too little about the paradoxical effects of warming and cooling, however, to tell what the outcome might be.

More sudden, perhaps much briefer but nevertheless disastrous, effects are predictable if chemical or physical means were developed for attacking one of the natural constituents of the atmosphere—ozone. A low concentration of ozone (O_3, a rare molecular form of oxygen) in a layer between fifteen and fifty kilometers altitude has the utmost significance for life on land. It is responsible for absorbing the greater part of the ultraviolet from the sun. In mild doses, this radiation causes sunburn; if the full force of it were experienced at the surface, it would be fatal to all life —including farm crops and herds—that could not take shelter. The ozone is replenished daily, but a temporary "hole" in the ozone layer over a target area might be created by physical or chemical action. For example, ultraviolet at 250 millimicrons wave length decomposes ozone molecules, and ozone reacts readily with a wide range of materials.

At present, we can only tentatively speculate about modifying the short-wave radiation at its source, the sun. We have discovered major instabilities on the sun's surface that might be manipulated many years hence. In a solar flare, for example, 10^{10} megatons of energy are stored in distorted magnetic fields. With advanced techniques of launching rockets and setting off large explosions, we may sometime in the future learn to trigger these instabilities.

For the near future, however, modification will not be in the short-wave incoming radiation but in the long-wave outgoing radiation.

The usual schemes for modifying climate involve the manipulation of large ice fields. The persistence of these large ice fields is due to the cooling effects of the ice itself, both in reflecting (rather than absorbing) incoming short-wave radiation and in radiating heat at a higher rate than the usual ground cover. A commonly suggested means of climate modification involves thin layers of colored material spread on an icy surface, thus inhibiting both the reflection and radiation processes, melting the ice, and thereby altering the climate. Such a procedure presents obvious technical and logistic difficulties. For example, if one wished to create a surface coating of as little as one micron thickness to cover a square 1000 kilometers in size, the total material for this extremely thin coating would weigh a million tons or more, depending upon its density. So the proposals to dust from the air some of the globe's extended ice sheets are unrealistic and reflect a brute-force technique, taking no advantage of instabilities within the environment.

Although it may be technologically difficult to change an ice cap's surface character, and thus its thermal properties, it may be possible to move the ice, taking into account the gravitational instability of ice caps. The gravitational potential energy of water as a thick, high ice cap is much greater than it would be at sea level. This fact makes it possible, at least in principle, to devise schemes for bringing about a redistribution in the ice. Indeed, A. T. Wilson has proposed a cyclical theory for the Ice Ages, based on this instability.

The main points of Wilson's theory are as follows:

1. Antarctica is covered by an ice sheet several kilo-

meters thick. Pressure at the bottom of the ice is great enough to keep the ice at or near its melting point; water is an unusual material in that a pressure increase lowers rather than raises its melting point. An increase in thickness of the ice sheet could result in melting at the bottom. The resulting ice-water mixture along the sole of the glacier would permit flow by a process of freezing and melting—a flow process much more effective than ordinary plastic flow.

2. If such an instability occurs, the ice sheet will flow out onto the surrounding sea, and a large ice shelf will be formed between Antarctica and the ocean around it. As a consequence, short-wave solar radiation will be reflected, and there will be enhanced loss of heat by radiation at the long wave lengths, causing cooling and the inducement of world-wide glaciation.

3. Once the ice shelf is in the ocean, it will begin to melt and eventually will be removed. The ice remaining on land will be much thinner than before. As the reflectivity of the southern hemisphere decreases with the melting of the Antarctic ice cap, the global climate will grow warmer again, corresponding to the start of an interglacial period. The ice cap will slowly form again.

Commenting on Wilson's theory, J. T. Hollin has noted the possibility of a catastrophic surge or advance of the ice sheet, such as has been recorded from small glaciers on numerous occasions. The largest surge yet reported is probably that of the ice cap in Spitsbergen, which advanced up to twenty-one kilometers on a front of thirty kilometers sometime between 1935 and 1938. There are also reports of glacial advances at speeds up to one hundred meters per day. Hollin speculates that, once the bottom-melting phase of a gravitationally unstable ice cap is reached, it will move quickly. In addition to trapped geothermal heat melting the ice at the bottom, there are additional contributions from frictional heat generated as the glacier scrapes along the solid ground.

If the speculative theory of Wilson is correct (and there are many attractive features to it), then a mechanism does exist for catastrophically altering the earth's climate. The release of thermal energy, perhaps through nuclear explosions along the base of an ice sheet, could initiate outward sliding of the ice sheet which would then be sustained by gravitational energy. One megaton of energy is sufficient to melt about 100 million tons of ice. One hundred megatons of energy would convert 0.1 cm. of ice into a thin layer of water covering the entire Antarctic ice cap. Lesser amounts of energy suitably placed could undoubtedly initiate the outward flow of the ice.

What would be the consequences of such an operation? The immediate effect of this vast quantity of ice surging into the water, if velocities of one hundred meters per day are appropriate, would be to create massive tsunamis (tidal waves) that would completely wreck coastal regions even in the Northern Hemisphere. There would then follow marked changes in climate brought about by the suddenly changed reflectivity of the earth. At a rate of one hundred meters per day, the center of the ice sheet would reach the land's edge in forty years.

Who would stand to benefit from such application? The logical candidate would be a landlocked equatorial country. An extended glacial period would insure near-Arctic conditions over much of the temperate zone, but temperate climate with abundant rainfall would be the rule in the present tropical regions.

Future of Weather and Climate Modification

The foregoing perhaps represents a more positive view of weather and climate modification than that held by many earth scientists. I believe this view is justified as it is based on three scientific and technological advances. First, understanding of basic meteorology has advanced to such an extent that mathematical models of the atmosphere here have been developed incorporating the most important elements. Physical processes in clouds, in turbulent exchanges at the surface, and in transmission of radiation through the atmosphere are no longer as mysterious as they once were. The volumes simulated by the models range from the size of a single cloud to the entire atmosphere; these models are no longer primitive representations.

Secondly, the advent of high-speed computers enables atmospheric models to be studied in greater detail. These computers have a peculiar importance to weather modification, since they will enable scientists to carry out extended experiments to test whether or not various schemes for manipulating the atmosphere are indeed possible and what the outcome should be.

The third advance lending support to expectations for weather and climate modification is the new array of instruments developed to observe and detect changes in the atmosphere. The most dramatic and perhaps the most powerful is the meteorological satellite, which provides a platform whence the atmosphere can be observed, not only in geographically inaccessible regions, but also with entirely new physical measurements. For example, meteorological satellites of the future will permit the determination of humidity, temperature, and pressure as averaged over substantial volumes of the atmosphere, providing quantities that are needed to develop the mathematical models. Sophisticated surface instrumentation, for observing detailed processes within smaller parts of the atmosphere, provides us with far more powerful tools with which to look at clouds and at the interaction of the atmosphere with its boundaries than those which were available ten or twenty years ago.

Earthquake Modification

What causes earthquakes? Over geological time, the irregular distribution of heat-producing radioactive elements in the rock layers gives rise to subsurface temperature differences between various parts of the earth. In the continents, granites and similar rocks have concentrated radioactive elements near the surface; no similar concentration has taken place in the suboceanic regions, which may as a result be more than one hundred degrees centigrade cooler than the corresponding subcontinental regions. Such variations in temperature along a horizontal line, due to the differences in the vertical distribution of heat-producing elements, give rise to large thermal stresses, causing strain analogous to that which cracks a glass tumbler filled with hot water. The strain tends to be greatest in regions of abrupt temperature change along a horizontal line through the earth's crust. The strain may be partially relieved by the slow convective flow of material in the deep earth which is thought by some geophysicists to push continents about. But the strain can also be relieved by sharp fractures or by movements along previous faults in rocks near the surface. Movement along a fault radiates energy outward, which results in an earthquake. Each year approximately 200 megatons of strain energy is released in this fashion, the largest earthquakes corresponding to energy of the order of 100 megatons. The energy released depends on the volume of material affected. The largest earthquakes take place along faults having a linear dimension of 1000 kilometers, whereas smaller ones take place along faults of one kilometer or less.

Major earthquakes tend to be located along two main belts. One belt, along which about eighty-five per cent of the total energy is released, passes around the Pacific and affects countries whose coastlines border this ocean, for example Japan and the west coast of North America. The second belt passes through the Mediterranean regions eastward through Asia and joins the first belt in Indonesia. Along these two belts, large earthquakes occur with varying frequencies. In California a large earthquake might be expected once every fifty to one hundred years, while Chile might expect such a disturbance once every ten to twenty years. Sometimes major earthquakes have occurred in regions ordinarily thought of as being free from risk. For example, the New Madrid earthquake of 1811-1812 devastated a large area of central North America but had only slight cultural effects because of the area's sparse population.

Today, our detailed understanding of the mechanism that causes an earthquake and of how the related instabilities can be triggered is limited. Only within the last few years have serious discussions of earthquake prediction begun, whereas moderately reliable weather forecasts have been available for about the last thirty to fifty years. Currently, substantial effort is being made, primarily by Japan and the United States, to develop techniques for forecasting earthquakes. These techniques are based to a large extent on the determination of changing strain conditions of materials in the rocks surrounding recognized fault zones. Of possible value is the observation that before an earthquake the accumulating strain accelerates.

Control of earthquakes is a prospect even more distant than that of forecasting, although two techniques have been suggested through recent experience.

1. In the course of the underground testing of nuclear

weapons at the Nevada test site, it was observed that an explosion apparently released local strain in the earth. The hypothesis is that the swift build-up of strain due to the sudden release of energy in an explosion discharges strain energy over a large volume of material.

2. Another method of releasing strain energy has appeared from pumping of underground water in the vicinity of Denver, Colorado, which has led to a series of small earthquakes. The hypothesis here is that underground water has provided local lubrication permitting adjacent blocks to slip by one another.

The use as a weapon system of the strain energy instability within the solid earth requires an effective triggering mechanism. A scheme for pumping water seems clumsy and easily detectable. On the other hand, if the strain pattern in the crust can be accurately determined, the phased or timed release of energy from smaller faults, designed to trigger a large fault at some distance, could be contemplated. This timed release could be activated through small explosions and thus it might be possible to use this release of energy stored in small faults at some distance from a major fault to trigger that major fault. For example, the San Andreas fault zone, passing near Los Angeles and San Francisco, is part of the great earthquake belt surrounding the Pacific. Good knowledge of the strain within this belt might permit the setting off of the San Andreas zone by timed explosions in the China Sea and Philippine Sea. In contrast with certain meteorological operations, it would seem rather unlikely that such an attack could be carried out covertly under the guise of natural earthquakes.

Modification of Oceans

We are still in the very early stages of developing the theory and techniques for predicting the state of the oceans. In the past two decades methods have been devised for the prediction of surface waves and surface wind distribution. A warning system for the tsunamis (tidal waves) produced by earthquakes has also been developed.

Certain currents within the oceans have been identified, but we do not yet know what the variable components are; that is, what the weather within the ocean is. Thus we have not been able to identify any instabilities within the oceanic circulation that might be easily manipulated. As in the case of the solid earth, we can only speculate tentatively about how oceanic processes might be controlled.

One instability offering potential as a future weapon system is that associated with tsunamis. These frequently originate from the slumping into the deep ocean of loosely consolidated sediments and rocks perched on the continental shelf. Movement of these sediments can trigger the release of vast quantities of gravitational energy, part of which is converted in the motion of the tsunami. For example if, along a 1000-kilometer edge of a continental shelf, a block 100 meters deep and ten kilometers wide were dropped a distance of 100 meters, about 100 megatons of energy would be released. This release would be catastrophic to any coastal nation. How could it be achieved? A series of phased explosions, perhaps setting off natural earthquakes, would be a most effective way. I could even speculate on planning a guided tidal wave, where guidance is achieved by correctly shaping the source which releases energy.

Brain Waves around the World?

At heights of forty to fifty kilometers above the earth's surface substantial numbers of charged particles are found which make this part of the atmosphere, the ionosphere, a good conductor of electricity. The rocks and oceans are also more conducting than the lower atmosphere. Thus, we live in an insulating atmosphere between two spherical conducting shells or, as the radio engineer would put it, in an earth-ionosphere cavity, or wave guide. Radio waves striking either conducting shell tend to be reflected back into the cavity, and this phenomenon is what makes conventional long-distance radio communication possible. Only recently, however, has there been any interest in natural electrical resonances within the earth-ionosphere wave guide. Like any such cavity, the earth-ionosphere wave guide will tend to sustain radio oscillation at certain frequencies in preference to others. These resonant frequencies are primarily determined by the size of the earth and the speed of light, but the properties of the ionosphere modify them to a certain extent. The lowest resonances begin at about eight cycles per second, far below the frequencies ordinarily used for radio communication. Because of their long wave length and small field strength, they are difficult to detect. Moreover, they die down quickly, within one sixteenth of a second or so; in engineering terms, the cavity has a short time constant.

The natural resonant oscillations are excited by lightning strokes, cloud-to-ground strokes being a much more efficient source than horizontal cloud-to-cloud discharges. On the average, about one hundred lightning strokes occur each second (primarily concentrated in the equatorial regions), so that normally about six lightning flashes are available to introduce energy before a particular oscillation dies down. A typical oscillation's field strength is of the order of 0.3 millivolts per meter.

The power of the oscillations varies geographically. For example, for a source located on the equator in Brazil the maximum intensity of the oscillation is near the source and at the opposite side of the earth (around Indonesia). The intensity is lower in intermediate regions and toward the poles.

One can imagine several ways in which to increase the intensity of such electrical oscillations. The number of lightning strokes per second could be enhanced by artificially increasing their original number. Substantial progress has been made in the understanding of the physics of lightning and of how it might be controlled. The natural oscillations are excited by randomly occurring strokes. The excitation of timed strokes would enhance the efficiency with which energy is injected into an oscillation. Furthermore, the time constant of the oscillation would be doubled by a fourfold increase in the electrical conductivity of the ionosphere, so that any scheme for enhancing that conductivity (for example, by injecting readily ionized vapor) lowers the energy losses and lengthens the time constant, which would permit a greater number of phased lightning strokes before the decay of an oscillation.

The enhanced low-frequency electrical oscillations in the earth-ionosphere cavity relate to possible weapons systems through a little-understood aspect of brain physiology. Electrical activity in the brain is concentrated at certain frequencies, some of it extremely slow, a little around five cycles per second, and very conspicuous activity (the so-called alpha rhythm) around ten cycles per second. Some experiments have been done in the use of a flickering light to pull the brain's alpha rhythm into unnatural synchrony with it; the visual stimulation leads to electrical stimulation. There has also been work on direct electrical driving of the brain. In experiments discussed by Norbert Wiener, a sheet of tin is suspended from the ceiling and connected to a generator working at ten cycles per second.

With large field strengths of one or two volts per centimeter oscillating at the alpha-rhythm frequency, decidedly unpleasant sensations are noted by human subjects.

The Brain Research Institute of the University of California is investigating the effect of weak oscillating fields on human behavior. The field strengths in these experiments are of the order of a few hundredths of a volt per centimeter. Subjects show small but measurable degradation in performance when exposed to oscillating fields for periods of up to fifteen minutes.

The field strengths in these experiments are still much stronger, by a factor of about 1000, than the observed natural oscillations in the earth-ionosphere cavity. However, as previously noted, the intensity of the natural fluctuations could be increased substantially and in principle could be maintained for a long time, as tropical thunderstorms are always available for manipulation. The proper geographical location of the source of lightning, coupled with accurately timed, artificially excited strokes, could lead to a pattern of oscillations that produced relatively high power levels over certain regions of the earth and substantially lower levels over other regions. In this way, one could develop a system that would seriously impair brain performance in very large populations in selected regions over an extended period.

The scheme I have suggested is admittedly far-fetched, but I have used it to indicate the rather subtle connections between variations in man's environmental conditions and his behavior. Perturbation of the environment can produce changes in behavior patterns. Since our understanding of both behavioral and environmental manipulation is rudimentary, schemes of behavioral alteration on the surface seem unrealistic. No matter how deeply disturbing the thought of using the environment to manipulate behavior for national advantage is to some, the technology permitting such use will very probably develop within the next few decades.

Secret War and Changing Relationships

Deficiencies both in the basic understanding of the physical processes in the environment and in the technology of environmental change make it highly unlikely that environmental modification will be an attractive weapon system in any direct military confrontation in the near future. Man already possesses highly effective tools for destruction. Eventually, however, means other than open warfare may be used to secure national advantage. As economic competition among many advanced nations heightens, it may be to a country's advantage to ensure a peaceful natural environment for itself and a disturbed environment for its competitors. Operations producing such conditions might be carried out covertly, since nature's great irregularity permits storms, floods, droughts, earthquakes, and tidal waves to be viewed as unusual but not unexpected. Such a "secret war" need never be declared or even known by the affected populations. It could go on for years with only the security forces involved being aware of it. The years of drought and storm would be attributed to unkindly nature, and only after a nation was thoroughly drained would an armed takeover be attempted.

In addition to their covert nature, a feature common to several modification schemes is their ability to affect the earth as a whole. The environment knows no political boundaries; it is independent of the institutions based on geography, and the effects of modification can be projected from any one point to any other on the earth. Because environmental modification may be a dominant feature of future world decades, there is concern that this incipient technology is in total conflict with many of the traditional geographical and political units and concepts.

Political, legal, economic, and sociological consequences of deliberate environmental modification, even for peaceful purposes, will be of such complexity that perhaps all our present involvements in nuclear affairs will seem simple. Our understanding of basic environmental science and technology is primitive, but still more primitive are our notions of the proper political forms and procedures to deal with the consequences of modification. All experience shows that less significant technological changes than environmental control finally transform political and social relationships. Experience also shows that these transformations are not necessarily predictable, and that guesses we might make now, based on precedent, are likely to be quite wrong. It would seem, however, that these nonscientific, nontechnological problems are of such magnitude that they deserve consideration by serious students throughout the world if society is to live comfortably in a controlled environment.

AUTHOR'S NOTE: In the section on weather modification I have drawn heavily on *Weather and Climate Modification* (National Academy of Sciences, National Research Council, Washington, 1966). A. T. Wilson's paper on "Origin of Ice Ages" appeared in *Nature*, vol. 201, pp. 147-49 (1964), and J. T. Hollin's comments in vol. 208, pp. 12-16 (1965). Release of tectonic strain by underground nuclear explosion was reported by F. Press and C. Archambeau in *Journal of Geophysical Research*, vol. 67, pp. 337-43 (1962), and man-made earthquakes in Denver by D. Evans in *Geotimes*, vol. 10, pp. 11-17. I am grateful to J.

Homer and W. Ross Adey, of the Brain Research Institute of the University of California at Los Angeles, for information on the experimental investigation of the influence of magnetic fields on human behavior.

FEARS OF A
PSYCHOLOGIST

B Y

OTTO KLINEBERG
France

Professor Klineberg is director of the International Center for
Intergroup Relations within the École des Hautes Études at the
Sorbonne, Paris, where his teaching and research are primarily
concerned with interactions between ethnic groups. He is author
of The Human Dimension in International Relations.

THE MATERIAL presented in this volume is terrifying. We
have for a long time faced the awesome statistics of "overkill" made possible by the development of nuclear weapons; now we are presented with a bewildering variety of
equally effective—or even more effective—techniques of
destruction. We have been impatiently, and so far vainly,
hoping for some results from the seemingly endless rounds
of negotiations on disarmament, or at least for a treaty on
nonproliferation of the bomb; now we learn that even suc-

cess in this respect would still leave the door wide open to other deadly alternatives. Since fear is a natural reaction to danger, we have reason, all of us, to be terribly, mortally afraid.

As a psychologist, involved together with all my colleagues in the attempt to understand human behavior, concerned with problems of mental health and human relations, committed to the application of psychological knowledge to international affairs, I have special reasons for my fears. It is my task as a psychologist to ask what the threat implied in the development of new weapons means to individual human beings; to their attitudes, their values, their goals; to the characteristics of their "human nature." The prospect is not a happy one.

Some psychological problems are specifically raised in the preceding chapters. The so-called psychic poisons, capable of inducing temporary or even permanent insanity in whole populations, rivaling in their potential horror the maddest dreams of the maddest scientists of fiction, have a special repulsion for those of us who have had some contact with the world of the mentally ill. The development of automatic weapons controlled by computers, of potential attacks by robots on civilian populations, appear to create the possibility of wars conducted on the basis of electronic judgment, even less capable of rationality than the human variety.

Dehumanization and Irresponsibility

We are confronted here with a process that psychologists and psychiatrists have labeled *dehumanization*. It is by no means new, but it has been increasing in range and intensity; it takes many forms. One of its most common and far-reaching aspects is the conviction that *they* are not human in the same sense that we are; we therefore have no need to treat them as if they were like us, or to be concerned with what they might be suffering if they were truly human beings. Throughout recent history racism of one kind or another has been the ally and support of dehumanization in this sense; *they* who are less than human may differ from us in inherited physical type (race) or in religion or language or national identity or ideology, or in any combination of these. The result is the same; we deny them a true humanity.

What I fear, and what I regard as highly probable, is that the robotizing of our weapons will greatly accentuate this process; one form of dehumanization will feed the other. The farther we, individual human beings, are from our target, the less we will be concerned about how others suffer under attack. Even under present conditions of conventional warfare, only a fraction of our military personnel comes into direct contact with the enemy; that is one reason why I have never been persuaded that personal aggressiveness played an important role as a motive for war. With these new weapons, however, the fraction will diminish even further; and there will be even less room for sympathy or empathy that might attenuate the suffering inflicted on others.

Another meaning to the concept of dehumanization has also been invoked, namely, that certain forms of warfare are contrary to human nature. This is a complex and difficult idea to apply in practice, because it assumes a knowledge of what kinds of military weaponry are "human" and what kinds "nonhuman" or "inhuman." I know of no body of scientific knowledge that would permit of a decision in this case. What is more "inhuman"—a submarine sinking, napalm, or sticking a bayonet through another man's belly? The human mind is a complex and fearful contradiction, capable of self-sacrifice and selfishness, of cooperation and conflict, of decency or destruction. What we consider "human" will vary with our culture, our religious beliefs, our philosophy of life. "Human" and "humane" look and sound very similar, but they do not mean the same. It remains true that we react with greater revulsion to the use of some weapons than others; it is perhaps my own *deformation professionelle* that makes me feel a special horror of "psychic" weapons that make men mad.

Not only do I fear the effect of the new weapons on human minds as well as bodies; I fear also the process and the people involved in the decision as to when and how such weapons are to be used. With regard to nuclear weapons, the widespread anxiety about the possibility of war through accident, or through action by a single irresponsible person, has been tempered by a knowledge that safeguards have been introduced; we are informed, for example, that no one person, acting alone, can activate the bomb. The installation of teletypes (hot lines) capable of immediate communication between crucial capitals has a similar purpose. (This is cold comfort at the moment; as I write this, news is flashed that the Chinese have the H-bomb, and there is no very direct communication between Peking and Washington.)

There are at present five "nuclear powers." It has been estimated (by Sir John Cockcroft earlier in this volume) that as early as 1971 an additional seven nations will have the potential to join the "club"; he speculates that ten or more might follow suit in the 1980s and thirty (!) in the succeeding decade. As a psychologist, my fears attach themselves particularly to the question of what kinds of people will exercise power in these nations. The world has had its full quota of pathological leaders; one trembles at the thought of what Hitler would have done if some of these weapons had been at his disposal. Elsewhere I have written:

> Psychiatrists and clinical psychologists are trained to distinguish between normal and maladjusted persons. . . . There is therefore available in this area a high degree of technical competence that, if recognized and objectively applied, might have saved the world from the excesses of the Nazi regime, and might also prevent a recurrence of such excesses. . . . We have already reached the point, in many countries, of requesting a physician's report on the physical condition of a candidate for public office; we are not yet prepared to ask for a similar report on his psychological condition. (*The Human Dimension in International Relations*, New York: Holt, Rinehart and Winston, 1964, pp. 65–66.)

Since the reaction to this suggestion is usually a tolerant smile, I have no hope that it will soon be adopted. Pathological leadership is always a possibility, and unless some controls are instituted, the predictions of Sir John Cockcroft only add to my fears.

Nationalism and the New Weapons

Even within what may be regarded as the "normal" range of mental reactions, the danger is great. We are, most of us, at the mercy of a kind of perverted logic fed by an exaggerated nationalism that makes it impossible to look at our relations with other nations with any objectivity. We are right and *they* are wrong. Exactly the same behavior (feeding the hungry, giving aid to a government in danger, even destroying crops or people) is noble and just when we do it, devious and immoral on their part. We never engage in aggression; it is always *they* who started hostilities, or were preparing to attack so that we were obviously justified in defending ourselves. (From this subjective point of view it has been suggested that *all* wars are de-

hought about the dangers may somehow contribute a lit-
e to their reduction.

I have referred earlier to the contradictions that consti-
te human nature. To add one more, man has the capacity
reason as well as for irrationality. When Freud said,
Where id was, there shall ego be," he was thinking of the
sibility that our blind impulses (he would say, in-
cts) could be controlled, and at times replaced, by truer
tact with reality, by better use of reason. Even if the
d is not yet ready to take the logical step of relinquish-
some of its treasured national sovereignty in the inter-
f peace, it may still be prepared to take thought about
and means of reducing the threat of destruction. I
I may be forgiven if I end on a note of hope rather
f fear.

fensive!) Long ago Pascal wrote: Vérité en deçà des Pyré-
nées, erreur au delà (Truth on this side of the Pyrenees, error on the other). This kind of ethnocentric perception
prevails, whatever the nature of the mountains that separate one nation from others.

This is nothing new. What is new is the intensity that
characterizes nationalistic attitudes, so that the man who
seeks a larger view, who tries to see the world as others see
it, runs the risk of supreme condemnation as "unpatriotic."
We are witnessing an outburst, a proliferation of nations
and nationalism precisely at a moment in history when nationalistic attitudes should be considered anachronistic and
lethal. The development of new weapons adds to the danger by giving to a nation the conviction that it now possesses the means to destroy the "enemy." Whether or not
it will decide to do so will probably depend on its estimate
of what the enemy will be able to do in retaliation, but the
unrealistic nature of nationalistic psychology makes it
equally likely that there will be an exaggerated view of
one's own destructive capacity.

The growth of nationalistic attitudes is due to a variety
of historical and political factors that I am not competent
to identify or discuss, but there are also psychological aspects of considerable importance. Many observers of the
contemporary scene are struck by a kind of pessimism regarding the future that is characteristic of many young people all over the world. There is a decrease in what has been
called "future time perspective," the willingness to forgo a
present gratification in order to obtain greater rewards at a
later time. Many of these youths appear to be concerned
only with the here and now. This may be associated with a
feeling of rootlessness, of not belonging, of having nowhere to go, a feeling akin to what many sociologists have
called "anomie." This in turn is not very different from the

description by psychologists of a lack of a sense of *identity*, of uncertainty as to who and what one is, of unwillingness to accept the self, with its failings as well as its virtues. The search for identity continues, however, and may take the form of bizarre and even antisocial behavior, sanctioned and supported by the peer group, symbolized by styles in dress, hair, vocabulary, and recreational activities. It may also lead, however, to seeking and finding a refuge in national identity, precisely because personal identity is so unsatisfactory. One may not know where one is going, but the nation remains, strong, noble, invincible, and indestructible.

This analysis is admittedly speculative and is based on impressions rather than on scientific data. I have no way of knowing how widely it would apply; obviously many young people would not fit the pattern just described. What does seem well established is that personal insecurity and concern with status are associated with ethnocentricism and chauvinism, and it is reasonable to conclude that, other things being equal, whatever makes people more insecure at the same time creates a propitious climate for the more extreme forms of nationalism. The circle is now complete; nationalistic attitudes lead to the development and acquisition of weapons, new and old, which create uncertainty and add to the feeling of personal insecurity, which in turn contributes to the encouragement of still stronger nationalistic attitudes. My fear is that this mechanism will make it easier for the chauvinistic leader to find followers and to carry public opinion with him.

Fear and Hope

One further fear is related to what has been called the self-fulfilling prophecy. If we expect certain events to occur, if we predict that they will occur, then (other things

being equal) they are more likely to occ usually given is that of a rumor that a about to fail. This results in a rush of p to withdraw their deposits, and because pared to meet all the demands on shor does fail. Other things must be equal, lain's prediction of "peace in our time" filled.

The mechanism involved is actuall applies exclusively to those situatio actions play a relevant part. If we—l ers—regard a war as inevitable, and do nothing to prevent it, then our tion contribute to the likelihood o proliferation of weapons, new ar tute a self-fulfilling prophecy. T course, for defence, not for aggi aggression? Who is more aggress who puts a chip on his shoul knock it off, or the boy (or challenge? The very existenc the greatest temptation to thei

In this chapter I have wr no predictions. There is af between being afraid tha prophesying that it *will*. F panic, immobility, an abc blind emotion; it can al search for ways of escap avoid or overcome dang many of us are too apa fate to extricate us, or pect of "what will be that my own fears h

WE HAVE BEEN HERE
BEFORE

B Y

PHILIP NOEL-BAKER
United Kingdom

Mr. Noel-Baker is a Member of Parliament and Privy Councillor
who has devoted most of his life to international affairs and particularly to disarmament. Among his books are The Arms Race
(1958) and The Private Manufacture of Armaments (1936). He
was awarded the Nobel Prize for Peace in 1959.

IN PROPORTION as the armaments of each power increase, so do
they less and less fulfill the objects which the Governments have
set before themselves. . . . It appears evident that if this state of
things were prolonged, it would inevitably lead to the very cata-
clysm which it is designed to avert, and the horrors of which
make every thinking man shudder in advance.

> —CZAR NICHOLAS II OF RUSSIA *in his proposal*
> *for the first Hague Conference* (1898)

WE HAVE BEEN compelled to create a permanent armaments industry of vast proportions. . . . This conjuncture of an immense

military establishment and a large arms industry is new in American experience. The total influence—economic, political, even spiritual—is felt in every city, every State House, every office of the Federal Government.

We recognize the imperative need for this development. Yet we must not fail to comprehend its grave implications. . . . In the councils of government we must guard against the acquisition of unwarranted influence, whether sought or unsought, by the military-industrial complex. The potential for this disastrous rise of misplaced power exists and will persist. We must never let the weight of this combination endanger our liberties or democratic processes. We should take nothing for granted.

Disarmament, with mutual honor and confidence, is a continuing imperative. . . . Because this need is so sharp and so apparent I confess that I lay down my official responsibilities in this field with a definite sense of disappointment. As one who has witnessed the horror and sadness of war, as one who knows that another war could utterly destroy our civilization . . . I wish I could say that a lasting peace is in sight.

—President Eisenhower *on re-linquishing his office* (1961)

WE HAVE BEEN HERE BEFORE. Each of the "conventional" weapons, such as high explosive, the submarine, and the airplane, was in its day a novel and frightening increment to the horrors of warfare. The scientists contributing to this book give a clear warning of yet more weapons, under development or conjectured, which will add to the capacity of human begins for killing one another.

The reaction of any reasonably minded reader must be a wish to see these further perversions of science stillborn. It is widely recognized that the only plausible way to insure that outcome is by general disarmament and the enforcement of international law. Again, we have been here before.

The end of war became overdue many decades ago, long before the H-bomb was invented. Throughout this century, national leaders, military commanders who have seen enough of war to be revolted by it, experts in every branch of learning, have testified to the need to abolish war. Every few years statesmen and diplomats have gathered for disarmament conferences; on each occasion most of the participants have wished passionately for success, but they have failed to make more than limited advances, quickly reversed by the march of events.

The new developments in weapons make the task of securing disarmament more urgent than ever, and in some technical respects more difficult, but the basic political issues are not new. This chapter deals with the impediments to disarmament and with some current international trends affecting the prospects for the future. Some of these trends are unfavorable, but some are encouraging.

The Militarists

When the peace-seeking policies of governments, reflecting the heartfelt wishes of their peoples, are consistently frustrated, it is reasonable to seek a powerful and persistent reason. Sufficient reason can be found in those influential individuals and groups whose political convictions, professional interests, or commercial instincts set them resolutely against disarmament. They are the militarists in our midst. Few of them are prominent on the political stage, for the very good reason that to favor war is to be politically unpopular. They are, as Lord (Robert) Cecil said in a letter to President Wilson in 1916, "able and honourable men"; but they are "past masters in obstruction," and they are very powerful, as has been proved by the history of the repeated failures to achieve disarmament and avoid war.

The militarists have been much the same kinds of people, at least for the last hundred years. They have included military staffs who are deeply imbued with the belief that wars are inevitable, and that "human nature does not change." Another group has comprised the armaments manufacturers and salesmen who, encouraged by their national governments to accept a "patriotic duty," develop a vested interest in preparation for war and in promoting the export of arms. Thirdly, there have been the patriotic societies, dedicated to the cultivation of nationalistic ideals and the glorification of military strength. Again, the influence of technical and trade journals, with editorials favoring military application of industrial resources, must not be underestimated. The activities of these groups interlock to create "the military-industrial complex" of which President Eisenhower warned his fellow countrymen.

There is an additional militarist class, most sinister of all: the secret services and intelligence agencies, which operate outside normal governmental control, isolated by their secrecy and left to pursue what they regard as their nations' best interests by underhand means. The mentality necessary to carry out such work in peacetime ensures that these agencies will attract at least a proportion of men for whom all foreigners are enemies and all liberal thought an anathema. Most work of secret agents remains forever secret, but that which does come to the surface is plainly seen to be inimical to the cause of disarmament and peace. If one singles out the C.I.A. in this connection, it is only because conscientious Americans have brought some of its activities to light in recent years; other nations, including my own, have their secret services, just as they have their "military-industrial complexes."

The Arms Race

The modern arms race may be said to have begun with the Franco-Prussian War of 1870. In that war, to the astonishment of the military experts, the Prussian conscripts totally and swiftly defeated the long-term professional army of Napoleon III. Napoleon had boasted that his troops were ready for battle down to the last gaiter button. But the discipline, the skill, and the courage of the Prussian conscripts were not inferior to those of the French professionals; while the steel guns with which Krupp had furnished them were notably superior to the iron guns of Schneider.

When that war was over, many governments decided that they must adopt conscription and re-equip their forces with modern arms. Thus the number of men in the national forces of the countries of the world greatly increased; the market for all kinds of military equipment expanded. In the last four decades before World War I, no industry in the world grew so fast as the production of arms, and no other investment held such a glittering prospect of quick and large returns.

Serious efforts to prevent arms races and secure international order can be traced back to the shock of the Franco-Prussian War, which gave some hints of what modern war would be like. But when the nations gathered at The Hague in 1899, on the initiative of the Czar of Russia, they were diverted from the serious discussion of general disarmament to what is now called "arms control." They produced some "laws of war," intended to preserve a gentlemanly atmosphere in battle. These "laws," prohibiting the use of gas, fire, air bombardment, and merciless attacks by submarines on merchant shipping, did not outlast World War I.

Nevertheless, in the years before 1914 many national leaders tried to resist the pressure for armament expansion. The chief cause of dissension between Germany and Britain was the battleship building programs reinforced by misinformation about the scale of the programs. At the eleventh hour, Britain offered a "naval holiday," but it was refused and everyone agreed that war was inevitable. Here we can identify von Tirpitz as the militarist who won the day despite the opposition of the German Chancellor and other leaders. But that does not exonerate the other militarists, on both sides. For years they had sustained frenzied and untruthful propaganda in support of battleship construction, starting with British construction directed against France in 1893. The competition was given a new twist by a British innovation, the Dreadnought, which rendered obsolete the previously existing battle fleets of both Britain and Germany. Ironically, the battleships played very little part in the war when it came, and there was only one inconclusive engagement between the main fleets.

The horrors of land warfare in 1914-1918 aroused deep revulsion and a passionate desire for world order. The League of Nations, for its first decade, was an imposing and successful start in this direction. For a while, militarism was firmly checked. But the economic disasters of the great slump, starting in 1929, brought about the downfall of some democratic governments and their replacement by militarist dictators, including Hitler. The Disarmament Conference of 1932-1933 started off with excellent intentions on all sides, amid demonstrations and declarations of popular support unparalleled before or since. When it looked as though it would succeed, the conference was systematically sabotaged by the militarists, and eventually it was allowed to fail. That failure brought down behind it the only major disarmament success between the wars, the

Washington Naval Treaty of 1922. This Treaty is sometimes described as a failure. In fact, it ended an intense
naval race between Britain, the United States, and Japan;
saved great sums of money; and gave Japan ten years of liberal, antimilitarist government, during which period a general disarmament treaty might have been made.

In 1936 the Western governments betrayed the Cove-
nant over Abyssinia, and, with Hitler in power in Germany,
Winston Churchill could predict that general war would
follow within three years. After his prophecy was fulfilled,
and an even worse war had been fought, the peacemakers
tried again.

The United Nations is in many respects a real advance
on the League of Nations. It benefits, particularly, from the
international work of a group of powerful agencies associated with it (of which more later). On some important
occasions it has shown itself capable of stopping or preventing war, and of long-term peacekeeping operations. The
standing in world politics of Dag Hammarskjöld and
U Thant, as Secretaries General, has been higher than that of
any national minister of foreign affairs.

Despite these favorable circumstances, the only measures
of disarmament achieved in twenty-two years after Hiroshima were those banning nuclear weapons tests in the atmosphere and prohibiting certain military activities in
Antarctica and outer space. These measures have not pre-
vented the French and Chinese from exploding bombs in
the atmosphere, nor have they stopped the development of
military space systems.

Long discussions about more comprehensive disarmament programs have faltered whenever they seemed to be
within striking distance of agreement. To rehearse all the
attempts and frustrations would be tedious because they
have been so numerous. The two greatest failures were in

1946 and 1955. At the first the Russians rejected out of hand, with extravagant attacks, generous American proposals for putting atomic energy under international control. In 1955 after three years' urging by the Western powers, the Soviet Union accepted their comprehensive proposals for disarmament. Thereupon the United States withdrew the proposals.

Militarism Today

The fact is that the world has not, since 1945, enjoyed even the brief respite from militarism that occurred after World War I. Instead, we have had a continuous arms race between the major powers, involving particularly nuclear weapons and delivery systems, and lesser arms races in troubled regions, particularly the Middle East and southern Asia, where the armament industries of East and West have found ready markets for their products.

Organized opposition to disarmament, most visible in the United States, but occurring also in the Soviet Union, Britain, France, and China, has triumphed. It is not difficult to see how these resolute minorities succeed. In the first place, comprehensive disarmament depends upon simultaneous good will from all countries in the negotiations; it is necessary only that the militarists secure the ascendancy in one major country to frustrate the discussions. Secondly, every failure and every increment in armaments adds fuel to the militarist flame.

Today, militarism assumes subtler guises than those of the bemedaled general gloating over his nuclear weapons, or the mad scientist of the horror film enthusiastically concocting new terrors for mankind; although it cannot be denied that some individuals fit these parts quite well. More typical is the bland civilian who reasons that a disarmed world would be more dangerous, that the "other side" will agree to disarmament only as a temporary move in the game, that nuclear war may not be as insupportable as is generally supposed. Sir Solly Zuckerman, until recently Britain's chief defence scientist, has condemned the fashion for "abstract" modes of thought about nuclear war, and for replacing strategic judgments by numerical calculations. He has denounced the argument that nuclear war is "feasible," and that organized society could survive it, as "unrealistic mumbo-jumbo."

Echoes from 1911 and 1935 are still heard in the advertisements of armaments manufacturers, in the editorials of the aerospace journals, in the impatient utterances of patriotic societies. For the most part, however, the militarists in our midst are smooth and sorrowful in speech, yet in their purposes more savage and (because more sophisticated) more sordid than Attila the Hun.

Prospects for Disarmament

Are the prospects for disarmament improving or deteriorating? It is at least possible—and necessary—to remain optimistic. International relations are not markedly better or worse than in the past; deterioration in some parts of the world is offset by miraculous improvement in others, notably in Europe. The great cost and rapid obsolescence of weapons, characteristic of the present age of advancing technology, provide governments with a strong motive for calling a halt. The limited steps in disarmament since 1945 demonstrate the possibility of agreement.

It is customary nowadays to regard China as the greatest impediment. Not only the Americans, but the Russians, too, are fearful of China's growing strength and may be unwilling to disarm unless China does the same. The assumption is that the Chinese will not even negotiate. It is a foolish assumption, because it has never been tested; on the contrary, China has been treated as a pariah among the nations. Before the recent domestic troubles, a high-ranking Chinese minister spoke privately of his willingness in principle to negotiate general disarmament with comprehensive inspection and control. The present internal strife is certainly not helpful, but it will presumably abate. Meanwhile, the United States and the Soviet Union are still primarily arming against one another.

"Difficult" Weapons

New weapons, especially biological and chemical weapons that require little space for manufacture or storage, entail special problems of inspection and control. The longer comprehensive disarmament is delayed, the more difficult will it be to cope with all the variety of novel systems discussed in this book. But it is quite wrong, and playing into the militarists' hands, to suppose that disarmament with control is made impossible by these innovations.

Biological weapons are probably the most difficult of all to control, but the United States Government assured the United Nations Disarmament Commission in 1952 that controlled disarmament was entirely possible in this field. It said the principal safeguards against biological weapons would be found "in an effective and continuous system of disclosure and verification of all armed forces and armaments." The character of the problem has not changed appreciably since that time.

The spirit in which the control of "difficult" weapons should be approached was well expressed by the American delegate in 1952:

It may be true that there is theoretically no foolproof safeguard which would prevent the concoction of some deadly germs in an apothecary's shop in the dark hours of the night. But when the United States proposes the establishment of safeguards to ensure the elimination of germ warfare along with the elimination of mass armed forces and all weapons adaptable to mass destruction, it demands what is possible and practical, and not the impossible. The United States is seeking action to insure effective and universal disarmament, not excuses for inaction [U.N. Disarmament Commission, 15 August 1952].

The aim in a control system is not to provide absolute certainty that all secret military activities will be detected, but to make detection so probable that no country will risk the political ignominy of being caught. Given disarmament, there is no reason why Camp Detrick, Porton, and other biological warfare laboratories should not be handed over, as valuable assets, to the international fraternity of biologists, to assist them, by cooperative effort, to wipe out preventable disease. Many other scientific laboratories are possibly self-policing, because there is usually a stream of foreign visitors, who would be quick to spot any suspicious activity. Laboratories engaged in secret work, nominally for commercial purposes, will be those of greatest interest to a control agency; unwillingness to open doors on request will itself provoke attention.

There are two additional safeguards, one existent and one hypothetical, that can reinforce any formal measures of inspection. The first is the network of friendships and common knowledge that unites scientists in all countries and which makes it difficult, in peacetime, for a scientist to change his work or "disappear" without the fact being widely known. As the contacts between scientists of all countries multiply, this network is becoming stronger. Secondly, the control agency could offer very substantial payments, and political asylum, to any individual who reports a violation of a disarmament treaty by his own country. Nor is it likely that such measures exhaust the possibilities. Scientific research has played an important part in the development of systems for detecting nuclear weapons tests; similar contributions can be expected to other forms of disarmament.

The International Society

"Transport is civilization," said Rudyard Kipling. Would that it were true! For, if it were, the world would be incomparably more civilized than in Kipling's time. Transport and telecommunications have wiped out barriers of time and space. Jet passenger aircraft cross the Atlantic in a few hours; communications satellites flash television pictures from continent to continent in a quarter of a second. But our ideological divisions remain. Jet aircraft, and rockets like those used for launching communications satellites, are also carrying bombs capable of annihilating whole cities.

Transport is not civilization, but it *is* society. It creates the links, the travel, the personal contacts, the exchange of goods, the investment of capital, the joint management of systems of communication, the intellectual and artistic bonds that make men in different places part of one community.

Transport has also made war and armaments out of date. By the later nineteenth century war was a disaster to all the parties, whatever its cause, whatever its outcome might chance to be. The Franco-Prussian War of 1870 and the Boer War of 1899 were only the most outstanding examples of this truth. In the close community of what the international lawyers had then begun to call the society of states, war and the competitive machinery of war were a historical anachronism.

There were statesmen in many countries in the nineteenth century who understood that what the society of states needed to promote the vital interests of its members was international institutions that would enable that society to live, like all other human societies, by the rule of an accepted and an all-embracing law. Atogether, from 1815 to 1900, about 200 disputes between states were referred to international arbitration—thirty of them, according to Sir Frederick Pollock, about "questions of title and boundaries"—that is to say, matters obviously of vital interest to the nations concerned.

In 1889 the Inter-Parliamentary Union was founded, expressly to popularize the idea of international arbitration. Léon Bourgeois, Prime Minister of France, led a powerful movement to transform the system of ad hoc arbitration into a standing Court of Law; and at the second Hague Conference in 1907 he had a partial success. The U.S. Secretary of State, William Jennings Bryan, thought it also important to create standing machinery for dealing with disputes in which international law did not provide a ready answer—what are now called "political" disputes. When World War I began he was negotiating Treaties of Compulsory Mediation, in which both he and others placed great hopes.

Technical Bonds between Nations

On a different level of international relations the governments of the world during the nineteenth century set up a large number of organizations to promote their common interests in matters of day-to-day importance. The Universal Postal Union was the most important and the most successful; it was followed by the Telegraphic Union, the Radio-telegraphic Union, the Metric Union, the International Institute of Agriculture, and by many more that dealt with subjects as diverse as railway freight transportation, public hygiene, literary and artistic property, the slave trade and the liquor traffic in Africa, the nomenclature of causes of death, submarine cables, the legal protection of workers, earthquakes, and commercial statistics.

The collective importance of these organizations was very considerable; and Leonard Woolf was right in arguing (*International Government*, 1916) that they constituted, within their respective spheres, a true beginning of international government. This was even more notably so in the International Commissions that were established to control the traffic on rivers which passed through more than one state, and to maintain navigation and other facilities on these rivers. The most important of these Commissions dealt with the Rhine, the Danube, the Scheldt, and the Congo. They had wide administrative, and even legislative, powers.

To these international institutions were added influential nongovernmental links between nations, notably through the international trade union movement and the Inter-Parliamentary Union, already mentioned. The International Polar Year of 1882-1883, in which thirty-five countries took part in concerted geophysical studies, was the forerunner of much greater collaborative scientific enterprises since then.

Between the two World Wars the League of Nations fostered impressive developments of constructive, technical intergovernmental work in many spheres. The United States, although not a member of the League, took part in, and subscribed to these activities; as a result its financial contributions to the League were second only to the British Empire's—an indication of the relative importance of these subsidiary activities.

Even more encouraging has been the post-1945 development of new international agencies associated with the United Nations: the Food and Agriculture Organization,

the World Health Organization, UNESCO, the World
Bank and the International Monetary Fund, and several
others. They have never had as much money as they really
need, but their influence is felt all over the world, and affects the everyday lives of everyone. They operate according
to international ideas and aims, without overmuch respect
for narrow national interests.

The control of nuclear fuels for peaceful purposes, the
regulation of broadcasting standards and frequencies, air-
safety regulations, vaccination regulations for travelers, machinery for the conclusion of tariff agreements (for example, the Kennedy Round)—all these and many other practical functions of government have been willingly surren-
dered to international institutions. The technical and
cultural requirements of our international society are
sweeping aside the barriers between nations in almost every
peaceful human activity, and eroding national sovereignty.

It is often said, even by many who earnestly wish for it,
that the concept of world government is idealistic and remote. That is patently absurd, because we already have
a great deal of world government. Woolf pointed out its
existence more than half a century ago, and since then it
has multiplied. We see it in action in the Security Council,
the U.N. General Assembly, the International Court of
Justice, and the specialized intergovernmental agencies.
The question is not *whether* world government is possible,
but simply how much we shall have, how strong we shall
make it.

The Futile Race

This interpretation does not underestimate the obstacles to further development of world government at the
political, legal, and military levels, least of all the continua-

tion of the arms race and of open conflict between nations. But it is pernicious to argue about whether disarmament must wait on the settlement of outstanding disputes between nations, or whether the perfection of our international institutions must precede or follow these other steps. All of these goals must be pursued simultaneously; success or failure in any one direction will affect progress toward the others. None is out of reach.

The modern arms race is nearly a hundred years old; we have traced it to the Franco-Prussian War of 1870. It paused halfway for a decade, in the 1920s. But, except for that decade, it has continued without respite. Allies and enemies have interchanged, nations have grown mighty and fallen, tens of millions have perished in war, but the armaments factories have never stopped.

Novel weapons, from the Maxim gun to the antiballistic missile, have created perpetual anxiety among the nations about "keeping a lead." The forecasts in this book show that there is no technological plateau on which the contestants can pause for breath, no ultimate weapon with which all will be satisfied. Hitherto, forecasts about new weapons have been largely a prerogative of the militarists, who say to their governments, "Faster! Faster!" May this book serve as an authoritative warning to the majority, and may the cry it evokes be "Stop!"

THE NEW WEAPONS

B Y

NIGEL CALDER
United Kingdom

WHY A SUMMARY? Why restate the principal conclusions of the expert contributors to this book? The reason is that the accumulation of possibilities and risks from many kinds of warlike activity deserves at least as much attention as the specific forecasts within each field of expertise.

The weapons discussed range from some that exist already, through others under development or plainly plausible, to some that may seem farfetched. Right or wrong, these last must represent the weapons of the more distant future about which we can only guess, knowing that unforeseeable discoveries or inventions are likely to generate even stranger military applications. The impression this book is intended to make, and this summary to reinforce, is that military technology is a Hydra: for each weapon that seems familiar and containable, others rise up threatening to defy containment; for every problem, generated by the military rivalries between nations, that attracts the attention of statesmen, others are looming scarcely noticed.

Even without the introduction of novel scientific principles or devices into warfare, technical improvement of "conventional" weapons, using projectiles, high explosives, and armor, is increasing their power to kill and to devastate. The chief reason is that the use of radar and other target sensors, of proximity fuses and of computers for fire control, greatly enhances the accuracy of each gun shot or missile round. Such techniques can probably deny a battlefield to infantry, deny the air to manned aircraft, and deny the sea to surface ships—unless countermeasures can frustrate the electronic systems. No one knows just how effective the competing techniques will be and, if a conventional "great war" were to break out between well-equipped nations in the future, it would be what General Beaufre calls a "truly enormous experiment." It could easily degenerate into bloody attrition worse than that of the two World Wars.

Such a major conventional war is in any case made improbable by the existence of nuclear weapons. Swift attacks, initiated by greatly superior forces and achieving their purpose within a few days, seem to be politically and militarily the only effective style remaining for conventional war. Against poorly equipped but well-organized guerrillas, on the other hand, the most sophisticated weapons systems may succeed only in "hitting air," and serve merely to postpone an inevitable political settlement.

Professor Dedijer takes the view that new weapons scarcely affect the principles of guerrilla warfare: guerrillas have always been at a disadvantage in firepower and they make it their business to be absent when a massive attack is launched against them. Dedijer characterizes guerrilla warfare as a politically motivated form of resistance in underdeveloped countries, nationalistic in nature and directed against foreign influences in the rule of the guerrilla's own country. If it matches the aspirations of the general population, it will tend to prosper and the only plausible strategy against well-organized guerrillas involves long and costly attrition in infantry engagements—destroying an idea by killing all those who hold it.

Here is the "poor man's power," the means by which underprivileged people can, when they feel driven to it, confront modern military forces with a good chance of success. With the poor of the underdeveloped countries tending to become poorer as populations explode, and with continuing interference by great powers in the affairs of nations within their "spheres of influence," the prospect is of an endless series of guerrilla wars, particularly in Latin America, Southeast Asia, and parts of Africa. The only way of averting such an appalling future, in Dedijer's view, is to remove the causes of social unrest, by massive aid and political reform. Neither in the scale nor in the manner of present aid is there much ground for hope.

The trend with which Sir John Cockcroft is chiefly concerned is the acquisition of nuclear weapons by countries not already possessing them—leading to greater likelihood of nuclear war breaking out, somewhere, between a pair of nations. The development of usable nuclear weapons and of their means of delivery is an expensive business, but within the resources of several nations, at least. For the manufacture of A-bombs, the cost of plutonium is really very low and the construction of nuclear power stations in many countries provides a ready-made source of nuclear explosive if international safeguards should fail or be disregarded.

Besides the existing nuclear powers, seven nations (Canada, West Germany, India, Italy, Japan, Spain, and Sweden) will have the potential to produce more than one hundred kilograms of plutonium per year, by 1971. Cockcroft expresses anxiety about a possible chain reaction, in which the acquisition of nuclear weapons by one new country would provoke other nations to follow suit. For example, if three nations made nuclear weapons for the first time in the 1970s, ten might do so in the 1980s, and thirty in the 1990s.

At present, the construction of H-bombs, rather than A-bombs, depends on uranium-235 as the triggering explosive, and the preparation of this material involves a very costly and cumbersome gas diffusion plant. As Dr. Inglis points out, the development of ways of obtaining uranium-235 more easily, or of using plutonium as a trigger, could greatly aggravate the problem of nuclear proliferation.

The laws of physics being what they are, Inglis believes with some confidence that there is no radically new principle for nuclear weapons to supplement the existing choice of fission and fusion bombs. Nor does he think that there is likely to be much extension of the choice of the materials used as nuclear explosives; for example, he dismisses the idea of a californium bullet as a "confusing fantasy." Existing nuclear weapons are as destructive as any military man could ask for. Variations and technical improvements are possible, of course, and even the hypothetical "doomsday machine," which would obliterate all life on earth, is not technically absurd, although it is almost certainly strategically absurd. A doomsday machine could plausibly be a series of extremely dirty H-bombs primed with cobalt. On the other hand, attempts will no doubt continue to make very "clean" H-bombs, which do not rely for detonation on the explosion of a fission bomb—fission products being the major source of fallout. But on the basis of present-day physics Inglis is sceptical about the practicability of a fission-free H-bomb.

The threat that Inglis sees comes not from new kinds of nuclear weapons but from a multiplication of the existing types of weapons, so that the world's nuclear arsenals will reach a fantastic level of destructive potential. Here he sees the chief risk in a U.S.-Soviet missile race, following the current development of tolerably practicable antiballistic missiles (ABMS). The anxiety is shared by Cockcroft, who points out that, at the present time when efforts are being made to limit the spread of nuclear weapons, a responsibility falls upon "nuclear powers" to curtail their own armaments.

The ABM will use a nuclear explosion to destroy an oncoming ballistic missile. Professor Stratton discusses the ABM from a technical viewpoint and emphasizes both the very great cost of even a partially effective system and also the very fast response it would need.

If developments in antiaircraft missiles go as Stratton thinks they may, manned strike aircraft will be kept so far from their targets that they may be effective only as carriers of "stand-off" bombs; in that case, the advantages of pilot judgment in tactical situations would be severely curtailed. The development of aircraft, missiles, and spacecraft for military purposes may be subject to important changes in guidance systems, using the laser. There may also be novel means of destruction; Stratton does not rule out the development of a laser "death ray," and he speculates about an explosive agent intermediate in force between high explosive and nuclear weapons.

The possibilities for satellites carrying H-bombs, for surveillance satellites, and for satellite interception, imply a serious risk of a complex and expensive arms race in space, involving unmanned satellites and manned space-craft. The fact that bombs parked in orbit will be very diffi-cult to intercept, once they have been "called down" to their targets, gives a strong incentive for the development of systems for inspecting and destroying hostile satellites while they remain in their predictable orbits.

Stratton is alarmed by the way automatic systems de-signed and programed by engineers may replace the politi-cal and military judgment of national leaders. He also em-phasizes the great scope for error and false assumption that exists in the complicated studies needed to choose aero-space weapons a decade in advance of their deployment in service.

The ways in which the use of computers in intelligence and control systems are revolutionizing the character of in-ternational relations and of war provide the main theme of Professor Wheeler's chapter. Already, the Commander in Chief of a superpower can—indeed, is required to—monitor events in all troubled areas of the world. Because of the speed and sensitivity of the systems, local incidents are known instantly, and decisions are made at the highest levels. One consequence is that swift military "solutions" come to seem more appropriate than old-fashioned diplo-macy. But in the future more refined systems will evolve, until not only are events known instantly but they can be anticipated by deduction from masses of computer-proc-essed intelligence data. The logical consequence will be pre-emptive strikes to forestall action by the adversary. The men who have to implement a decision to strike will, at such a stage, be simply worse informed than the computer system, so that they may be obliged to follow its proposals. It is in that sense that computers may come to govern hu-man affairs and lead us into war.

Besides the computer, another radical development in

human affairs during the remaining years of this century will be the exploitation, for civilian purposes, of the resources of the deep ocean. At the same time, developments in naval submersibles, including already existing nuclear weapons systems such as Polaris, are creating a new three-dimensional battleground of the greatest strategic importance. As Professor Nierenberg describes the trends, we must expect three concurrent hazards to peace: (1) attempts by nations to appropriate large volumes of the oceans for commercial or strategic purposes; (2) intensive development of undersea and antisubmarine weapons favoring or compromising the submarine-borne strategic missile systems; and (3) an intermingling of civilian and military activities, in the same stretches of ocean, which can only tend to generate friction and suspicion.

The naval technology envisaged includes both the development of operational craft capable of reaching the greatest ocean depths, and the deployment throughout the oceans of automatic sensors providing continuous surveillance of activities at all depths. Although the submarine and the surface-to-surface missile will be so effective in the future that the sea lanes will be denied to all conventional shipping, the balance will be restored by the advent of big surface-effect vessels related to the hovercraft. These will be capable of speeds approaching 200 kilometers per hour, both for carrying cargoes and for naval purposes, including antisubmarine operations. The future of the aircraft carrier is put in doubt, even in a surface-effect version, by vertical-takeoff aircraft and an "oceanographic" aircraft of extremely long endurance, which Nierenberg envisages. We must also expect the creation of bases, in fixed positions on the sea: floating air bases anchored at strategic points around the world and submarine bases sited far from land at the edge of the continental shelf.

An indirect contribution of the oceans to future arsenals is suggested by Professors Fetizon and Magat, who note that certain fishes, notably the Tora Fugu, produce deadly poison. The potency of chemical weapons is probably greatly underestimated in current thinking about future warfare. Even if all research had stopped—which it certainly has not—military forces would still have the extremely effective "nerve gases" developed by the Germans during World War II. Against these gases, which can be absorbed through the skin and which kill by the interruption of the natural nerve control of muscle, protection is very difficult for armies and virtually impossible for civilians.

Although lethal gases have not been used very widely since World War I, other forms of chemical warfare have been perfected. For incendiary purposes, napalm is likely to continue in service in the decades ahead. The Vietnam war has brought wide-spread use of antivegetation poisons developed from weed killers. For very large-scale destruction of vegetation or crops, heat from H-bombs is likely to be more effective; alternatively, chemical warfare in the future may include agents released by rocket in the ozone layer of the atmosphere which create a temporary "hole" in this layer. As the layer serves to protect terrestrial life from the intense ultraviolet radiation of the sun, its local obliteration, by chemical means, could result in fatal "sunburn" for all vegetation and exposed animals in the underlying region.

The chief current development in chemical weapons that Fetizon and Magat discuss concerns the "psychic poisons." Agents are already known that can have profound effects on the human mind, and new ones are being sought for medical purposes. LSD is the best known of the existing agents, but other types are also available for study. Typically, LSD given in controlled doses induces a state of temporary madness such as could make soldiers throw away their arms or sit down and weep like babies. Fetizon and Magat strongly challenge, however, the suggestion from enthusiasts for psychic weapons that here is a humane, non-lethal means of waging war. They point out that if most people in the target population are to receive effective doses, many will receive large overdoses, which will cause permanent insanity or death. Moreover, the military administration of such drugs to armies or civilian populations in the midst of their normal business will create horrific chaos in which many people will die.

Microbiological weapons have even more fearful consequences for human targets, possibly exceeding the killing power of all-out thermonuclear warfare. Professor Hedén describes how a cloud of infective microorganisms could strike down by disease the human inhabitants of a whole province, or ruin their crops. Such weapons can be released swiftly from spraying vehicles over large areas or covertly by saboteurs against selected targets such as military staffs, crews of ships, or big public assemblies.

Biological weapons are easy to make and they may be especially tempting for nations unable to develop nuclear striking forces. Small groups of individuals may be able to upset the strategic balance. On the other hand, advanced microbiology may evolve novel forms of disease for military purposes. Defence against biological attack will be peculiarly difficult—most of all in the developing countries, which lack good public health facilities.

Biological attack may, in practice, be indistinguishable from epidemics—and vice versa—so that allegations of biological attack may become frequent, and it will be hard to tell whether they are well founded. Reprisals can lead to an escalation of the intensity and virulence of the attacks. It will scarcely be possible to confine the diseases so evoked to the target areas. The very young, the very old, and the sick will be especially vulnerable to biological warfare.

The impact on weapons systems of electronics, and of compact computers in particular, may manifest itself in bizarre robots like the "walking bomb." Professor Thring forecasts the development of unmanned, expendable, and practically unstoppable tanks on legs, capable of finding their own way slowly but surely to their targets. Such a delivery system would introduce a novel psychological factor into the use of thermonuclear weapons, by allowing a rather long period for second thoughts between the launching of an attack and the first explosions.

More generally, robot foot soldiers, aircraft, and submersibles may displace men from the conduct of the battle once and for all, as such computer-controlled systems come to surpass men in tactical skill and reliability. Once a decision to exclude men has been taken, the traditional role of infantry units in controlling ground might reassert itself even in situations where men simply could not survive. Important limitations on the design and performance of aircraft and submersibles disappear when the crews' requirements for comfort and survival no longer figure in the calculations. By such developments some purposeless loss of life among servicemen may be avoided, but human populations will still be the targets of the robot strategic weapons.

The possibilities of geophysical warfare, aimed at producing subtle or catastrophic modifications in the condition of the earth or its atmosphere, are largely speculative. But it is important to understand that the impediments arise more from ignorance of natural processes, which leaves the long-term effects of particular actions incalculable, than from any basic incapacity for human interference with the environment. There is growing evidence that relatively modest action can trigger the release of much greater energy, or alter the course of a natural process. Professor Mac-Donald does not regard as impossible strategic geophysical action of various kinds which he discusses.

For example, economic attrition by drought might be brought about by systematic seeding of clouds in a prevailing airstream, to remove moisture. Hurricanes might be guided toward an opponent's coastline. Remote triggering of a major earthquake is not entirely incredible, nor is the creation of an artificial tsunami (tidal wave) by tipping loose material off the edge of the continental shelf. As an extreme form of geophysical warfare, one can imagine deliberate inauguration of a new Ice Age, by interference with the Antarctic ice cap.

MacDonald also points out that human mental performance may be subject to geophysical influences. Although he admits that his scheme for timing lightning flashes to disturb people on the other side of the world is far-fetched, he thinks it very probable that environmental means of manipulating behavior will emerge within the next few decades.

The trends in weapons development already carry psychological implications. Professor Klineberg expresses a personal revulsion for the "psychic poisons" but he notes that this may be professional bias of one who has been concerned with mental illness; no really objective definition of "human" or "inhuman" weapons seems possible. On the other hand, the automation of warfare and the possibilities for attack from great distances must tend to dehumanize military operations and whittle away any remaining sympathy for the sufferings of one's enemies.

Knowing of the range of human mental reactions, and of the world's past quota of pathological leaders, Klineberg would require that candidates for public office be subject to psychological examination, but he has no hope that this will soon be done. He fears that new Hitlers may come to control nuclear weapons or other new systems.

Klineberg also points to the association between personal feelings of insecurity and nationalism—the finding of security in national identity and its irrational corollaries of racism and chauvinism. But new, frightening weapons may add to personal insecurity, thus heightening nationalist feelings and encouraging the further development and manufacture of weapons. The circle is complete and promises a grim future for mankind, unless irrational fears and apathy can be replaced by constructive, reasoning action in response to fears that are all too well-grounded in the facts.

Mr. Noel-Baker testifies to the possibility of remaining hopeful that the nations will agree on general disarmament and thus come to restrain the application of science to destructive purposes. I shall not attempt to summarize his discussion of the interplay of militarists and peacemakers, of problems of disarmament created by new weapons, and of the growing technical ties between nations. His contribution is less concerned with the scientific and technological aspects of possible new weapons than with the phenomena of the arms race and rapid military innovation, which have a continuous history from 1870 to the present. I shall just close in the same spirit as his conclusion.

This book is not intended as a manual for weapon-makers. It is not even meant to provide an exhaustive list of all the possibilities for new weapons: there are other sciences from acoustics to zoology in which quite different possibilities could be discovered. As long as men want ingenious ways of killing or dominating one another, the natural world, through the medium of science, will provide them. The same ingenuity and knowledge of nature can be applied quite otherwise, to creating a healthy, pleasant, and exciting environment for all mankind. But the worst forebodings will surely be fulfilled, and even the most modest visions of a better world will be smashed, if present military tendencies continue. Our loss will be a double one—both of what we have and of what we might have made—unless peace comes.